D0854927

THE PATTERN OF EXPECTATION

THE PATTERN OF EXPECTATION

1644–2001

I. F. CLARKE

BASIC BOOKS, INC., PUBLISHERS
NEW YORK

For
Catherine,
Christopher, and Julian
who are the future

First published in the United States by Basic Books, Inc., New York 1979

Library of Congress Catalog Card Number: 78–19669

ISBN 0–465–05457–9

Printed in Great Britain

Contents

Illustrations

Between pages 52 and 53

The pictures not otherwise credited are from the author's private collection.

Acknowledgments

I am glad to have this opportunity to acknowledge the help and advice I have had from friends and colleagues.

I am especially grateful to the Librarian and staff of the Andersonian Library, the University of Strathclyde, for their unfailing assistance over many years; and in particular I wish to thank Audrey Calder of the Inter-Library Loans Service who was most resourceful in dealing with the very varied requests of long reading lists.

Several of my friends have devoted many hours to reading the various drafts of this book. Donald Gordon gave me the benefit of his considerable knowledge of philosophy; Professor Alan Sandison provided an invaluable commentary on the text; Jeremy Lewis, now of the Oxford University Press, helped in the early stages of planning the book; and Damazy Napiontek translated long passages from the Polish of Stanislaw Lem into English.

I thank in particular Margaret Dalrymple who was always equal to the many demands on her. She typed several drafts and presented a final version that was meticulous enough to please the editor, Margaret Stevenson, to whom I am also indebted for her painstaking work on the text.

Most of all I thank my wife, my most candid critic, who has had to live with this book for the last three years.

I.F.C.

The University of Strathclyde
1979

Prologue
The Idea of the Future

One of the more obsessive preoccupations of our time is the search for the true shape of things-to-come. In a period of exceptionally rapid change so much depends on the ability to foresee the direction in which society is moving. And so, since the end of the Second World War all the major industrial countries have appointed national commissions and specialist committees to investigate the pattern of future development. In response to this demand for reliable information about coming things the new profession of futurology sprang into existence in the 1950s; and during the last 10 years the journals of the futurologists have become required reading for politicians and planners from Sydney to Seattle.

The international conferences about the future, the planning boards and the technological forecasters all repeat the same message. Resources are limited, they say. Careful planning of future developments is essential. Many hopes for the twenty-first century will depend upon willing agreement and close co-operation between the peoples of our planet. So much has been said and written about the various social, political and technological trends at work throughout the world that the scenario of the future must be familiar reading for the majority of literate citizens everywhere. What was yesterday a familiar fantasy—atomic energy, space travel, lunar exploration—is today a commonplace reality. Indeed, according to a professor of physics at Princeton University, that dream of science fiction stories – the space colony – will be the achievement of the early twenty-first century, if the Subcommittee on Aerospace Technology of the United States Senate listens to what Professor O'Neill told them on January 19th, 1976. So, now that the discussion of the future has become a matter of universal interest, the time seems right to examine the idea of the future in all its major manifestations, from the earliest prophecies of the eighteenth century to the forecasts of doom and delight that nowadays fill the Sunday newspapers.

The professional horizon-watchers of our time are new arrivals in the brief history of prediction. Their forecasts mark the most recent stage in an evolutionary sequence that began with the anonymous account of a future British empire in *The Reign of George VI* in 1763 and with Sebastien Mercier's celebrated prophecy of an ideal world in *L'An 2440* of 1771. These utopian projections were the first major attempts to imagine the future circumstances of mankind; and their special interest for the modern reader lies in the entirely new attitude to time that they display. They reveal the dawning of a new awareness that the future will be different from the present. They are the beginnings of a vast new literature of expectation, which has been a primary characteristic of the industrialized nations ever since James Watt invented the separate condenser and set the modern world in motion.

During the last 200 years the constant advances in science and technology have transformed the condition of human life throughout our planet. The everyday experience of social change and industrial development has become so much part of our general thinking that we are apt to forget the extraordinary novelty of seeking to discern the shape of coming things. The frequent projections of tomorrow's world in films and television programmes hide the fact that for most of human history the image of the future has been a blank. Nowadays we know what was hidden from Roman emperors and medieval popes: that scientific discoveries and technological inventions can have the most far-reaching consequences for the world in the years ahead.

Throughout many centuries the evidence of the past seemed to show that the unchanging destiny of the human race was to repeat the eternal round of an agrarian civilization. The windmill, the waterwheel and the horse would continue for ever to provide their totally predictable amount of energy and motive power; and travellers could never expect any improvement in the rattling, springless coaches that took two weeks to make the journey from London to Edinburgh. Plagues, uncontrollable diseases and famines were the guarantee of small populations everywhere. Until the agricultural improvements of the eighteenth century and Jenner's discovery of the smallpox vaccine began to affect the even balance between life and death, it was to be expected that the majority of newborn children would not live past their fifth birthday.

But from the earliest times, wherever there was a knowledge of science, there was the hope that one day men would learn how to master the powers of nature. That was the conviction of Roger Bacon – medieval monk, philosopher and mathematician – who constantly asserted the utility of scientific studies. Some time about the year 1260 he wrote into the *Epistola de Secretis Operibus* his belief that the application of scientific knowledge could lead to the invention of the self-propelled ship, the motor car and the aeroplane:

> Machines for navigation can be made without rowers so that the largest ships on rivers or seas will be moved by a single man in charge with greater velocity than if they were full of men. Also cars can be made so that without animals they will move with unbelievable rapidity ... Also flying machines can be constructed so that a man sits in the midst of the machine revolving some engine by which artificial wings are made to beat the air like a flying bird.[1]

For close on 300 years such thoughts remained the private aspirations of exceptional minds. Then came the great intellectual advances of the sixteenth and seventeenth centuries which changed the way the Europeans saw themselves and their world. As the voyages of Columbus and Magellan revealed new lands and new peoples across the seas, the scientific investigations of Vesalius and Harvey, of Galileo and Kepler, opened up new areas of knowledge about man and his universe. All these discoveries encouraged a sense of expectation, which found an appropriate form of expression in the utopian fiction of the Renaissance. The device of the imaginary voyage, which St Thomas More invented in the *Utopia*, represented an expansion of the imagination. It enabled writers to describe their schemes for human happiness and social perfection in the mythical here-and-now of supposedly contemporaneous societies.

The hope of human happiness has haunted civilized man ever since Plato, our first father in utopia, outlined the geometry of the ideal state in the dialogues of *The Republic* and the Atlantis myth of the *Timaeus*. In a later age and a very different society St Thomas More took up that hope in the *Utopia*, and he brought the promise of perfection closer to reality by presenting his ideas as reported facts within the new accounts of geographical exploration. The description of the voyage to the far-off island, like the more recent journeys to some future time, is a potent literary device that reveals the variety and vigour of the motive forces at work in any society. In the beginning the image of the possible was the voyage across the boundless ocean to that other-world in which a happier history, or a higher morality, or the fortunate conjunction of political theory and scientific knowledge had established the ideal commonwealth.

For two-and-a-half centuries, from the *Utopia* in 1516 to the emergence of the first futuristic utopias in the late eighteenth century, the detailed description of the arrival in the unknown land – storms, wrecks, pirates and all – had served as prelude to the real business of exploring the range of possibilities in contemporary society. Out of this practice came the first rudimentary images of the future in European fiction; for the style of life in these ideal states belonged to the future in that their social and technological achievements were far beyond the capacity – but not the ambitions – of contemporary Europe. One example of this combination of new science and old morality is *La Città del Sole*, which the Dominican friar Tommaso Campanella began to write in 1602. The final version of 1623 described the ordered, virtuous and progressive society of the Solarians. They practise all the Christian precepts and they enjoy the benefits of an advanced technology. When they go to work in the fields, 'they use wagons fitted with sails which are borne along by the wind, even when it is contrary, by the marvellous contrivance of wheels within wheels.' The wish-fulfilment element comes out even more clearly in the account of the Solarian ships: 'They possess rafts and triremes, which go over the waters without rowers or the force of the wind but by a marvellous contrivance!'

Similar thoughts occupied the Lord Chancellor of England about the year 1620, when he was engaged on the final draft of *The New Atlantis*. In that renowned and most influential narrative Francis Bacon played the parts of prophet and preacher. Writing in the name of the new sciences, Bacon described a social system dedicated to the domination and exploitation of nature. The objective towards which all citizens are directed in the imaginary island of Bensalem is 'the knowledge of causes and secret motions of things, and the enlarging of the bounds of human empire to the effecting of all things possible.' Bacon shows that the pursuit of scientific knowledge had given his ideal commonwealth immense advantages – aeroplanes, submarines, new agricultural techniques, new means of warfare, advances in medical practice. At the centre of activity in Bensalem is Solomon's House, which co-ordinates and directs all the research work. As Bacon's secretary explained in his preface to *The New Atlantis*: 'This fable my Lord devised to the end that he might exhibit therein a model or description of a college instituted for the interpreting of nature and the producing of great and marvellous works for the benefit of men.'[2]

The fabulous element was a necessary constituent of the seventeenth-century utopias. Even at their most perceptive and original, as in *The New Atlantis*, the authors concerned themselves with aspirations rather than with anticipations. They were caught between an eagerness to demonstrate their theories of social improvement and an inability to foresee the consequences of the technological developments they so earnestly desired. Although the citizens of their ideal states were supposed to have achieved 'the enlarging of the bounds of human empire', they existed in the eternal paradise of a static society. The prodigious improvements in medicine and agriculture had not caused any increase in population; and there was never a hint that a highly advanced technology would give a community the mastery of the seventeenth-century world.

The terrestrial utopias were heroic essays in the unpredictable, for their authors could only assume that the progress of the sciences would increase the sum of human happiness. Their importance in their time, however, is beyond questioning. They mark the transition from the medieval preoccupation with the joys of an everlasting hereafter to those fascinations of an infinite progress of knowledge so typical of modern times. Their special value was to manifest a new sense of human destiny. The imaginary voyage to the ideal commonwealth was the perfect symbol of a great advance in human affairs. In their day they were the most complete expression of that belief in the powers of science so aptly summarized by Joseph Glanvill, Fellow of the Royal Society and Chaplain-in-Ordinary to Charles II:

> And I doubt not but posterity will find many things, that are now but Rumours, verified into practical Realities. It may be some ages hence, a voyage to the Southern unknown tracts, yea possibly to the Moon, will not be more strange than one to America. To them that come after us, it may be as ordinary to buy a pair of wings to fly into remotest Regions, as now a pair of Boots to ride a journey. And to confer at the distance of the Indies by Sympathetick conveyances may be as usual to future times as to us in a literary correspondence.[3]

The optimists had to wait until the second half of the eighteenth century for a clearer view of all the wonder that would be. In that still innocent age, when the first descriptions of time-to-come were beginning to establish a new prophetic literature in Europe, the rapid advances in mechanical engineering and in medical science were welcome signs that mankind was moving towards a very different kind of existence. The European response to the unprecedented changes of that time was to develop a new sense of perspective that discerned the shape of things-to-come in the circumstances of contemporary society. This unique understanding of the dynamic relationship between present and future is certainly the most distinctive and possibly the most extraordinary of all those changes – from balloons and steam-engines to atomic bombs and space laboratories – that separate the modern epoch from the rest of human history. It was an acquired aptitude of mind that found its own specific mode of communication in descriptions of future societies, accounts of future wars, and fantasies of future times.

Any examination of this engagement with the future will show that the apparently simple device of projecting the image of coming things is the subtle product of many original ideas. The first and the most powerful of these was the idea of progress, which encouraged the expectation of technological advances and social improvements in the eighteenth century. As the first great developments in the applied sciences gave proof that the world was beginning to change, the tale of the future appeared as the most suitable means of describing whatever a writer thought would happen in the centuries ahead. After the initial hesitations and uncertainties peculiar to an emergent literary form, the new fiction gained strength and authority from the growing awareness of the process of development at work in human history; and then, during the first half of the nineteenth century, the geological evidence for the immensity of time-past and the new evolutionary theories joined with the spectacular innovations in technology to provide a more extensive base for the tale of the future.

This all-embracing belief in continuing change was the Victorian doctrine of progress. As writers became more proficient in their handling of futuristic fiction and as the steady rise in the level of literacy made many more readers familiar with their extrapolatory visions, there was an increasing precision in the description of coming things. By the 1870s the tale of the future had become a most popular and very effective form of fiction. Ideas begot prophecies in a great outpouring of imaginary wars, utopias, scientific romances and tales of interplanetary adventure. At the same time, the rapid advances in technology and the many social changes promoted the first essays in forecasting the probable state of society in the next century – bigger cities, better communications, a world federation, even universal peace. For close on five decades this literature of the future continued to grow in volume and variety up to the outbreak of the First World War. And then the unforeseen and disastrous triumph of the military technologies started the debate about the consequences of scientific inventiveness that still goes on.

At the centre of this debate is the idea of the future – that estimate of the human capacity to change the world for better or for worse; that set of aspirations that would define the direction of future developments; the assessment of those novel conditions that could affect all living things throughout the planet. This long-continuing discussion of things-to-come has evolved through many forms, from the utopian prophecies of the Enlightenment to the most recent predictions from the institutes of futurology. How it was done, how the design of the picture has been cast and recast, is the subject of this book. And here it is necessary to add that, although this book will trace the development of futuristic fiction and of predictive literature, the central theme must always be the changing relationship between science and society during the last 200 years.

The record of that relationship is the history of modern urban civilization in search of a final but always elusive system of stable equilibrium. In the course of this advance from steam-engines to nuclear power stations, the tale of the future has developed an admirable means of picturing the ceaseless striving to adapt to the changing circumstances of a technological epoch. Indeed, the principal functions in the tale of the future correspond to major operations within society. Thus, theories about the reconstruction and renewal of the state have always found an appropriate form in those utopian schemes for human betterment that could be realized if only the world would attend to the proposals of Sebastien Mercier in *L'An 2440* or of Edward Bellamy in *Looking Backward*. The parallel ambition to expose the dangerous tendencies in society finds expression in nightmare projections that reveal, for example, the horrors of a ruthless plutocracy in Wells's *When the Sleeper Wakes* or the dilemmas of modern industrialism in Kurt Vonnegut's *Player Piano*. In a similar way, the perennial anxiety for the safety of the nation has an admirable means of communication in many tales of future warfare; and these developed into a popular and often notorious form of propaganda in the years between 1871 and 1914, when the Europeans fought out their differences in the imagined encounters of the *Zukunftskrieg*, *les guerres imaginaires*, and 'the story of the next great war'. And finally, the limitless opportunities of science fiction encourage the free play of the imagination in original stories of ingenious machines and advanced technological societies, marvellous adventures in future ages, fantasies of other worlds and other forms of life in distant galaxies.

The utopian impulse to anticipate the conquest of all the obstacles of time and space has always been a major element in futuristic fiction. It is only by virtue of the infinite liberty of the imagined future that a writer is able to range at will, unconstrained and godlike in his capacity to create new worlds in his own image. The realization of this Promethean imperative depends on literary conventions that have been developing ever since the stories of steam-driven balloons and electrical mowing-machines in the first decades of the nineteenth century. At every stage in the evolution of this fiction there has been an intimate connexion between the manner of the narratives and the matter of the prophecies. The earliest ideal states of the future, for example, invariably opened with elaborate explanations of the way in which the dreamer arrived in the twenty-fifth millennium or the author chanced upon his history of the future. Later on, after the idea of progress had become a dominant belief, the universal conviction of perpetual improvement gave writers the confidence to plunge directly into their accounts of imaginary epochs. This delicate balance between the expectations of an age and the presentation of futuristic fiction appears most clearly in the evolution of interplanetary themes—from the first primitive accounts of lunar journeys by balloon in the eighteenth century to the stories and films of heroic adventures in deep space that are so popular today. The world has delighted in the marvels of the Stanley Kubrick film, *2001: A Space Odyssey*, for the reasons that caused the editor of the *Journal des débats* to serialize Jules Verne's very successful account of a space journey, *From the Earth to the Moon*, in 1866. They are celebrations of the power and resources of a technological society; and they receive their initial validation from the ways by which the authors relate their themes to the contemporary understanding of scientific possibilities. In a more profound and subtle way, however, the ideas that power the tale of the future derive from the sources of hope and fear in human personality. Because there has always been this constant commerce between the mood of the day and the projections of imaginative writers, the history of futuristic literature displays a succession of vivid images; and these contrast, for example, the exuberant vigour of nineteenth-century optimism in Verne's account of the lunar journey with modern anxieties about the uses of science in the victory of the robots in Karel Capek's *R.U.R.*

The tale of the future tends to be a literature of extremes, since it is only possible to meet the essential requirements for these projections by tracing the curves of hope or fear to their logical conclusions in visions of social perfection, or in forecasts of terrible wars, or in extravagant fantasies of human power. Indeed, the contemporary interest in space fiction points to the subconscious depths from which these fantasies emerge. The desire to be saved by a visitation of better and wiser beings, like the parallel hope that the Western style of society will survive for ever, has become part of contemporary mythology. Today, without any effort of the mind, millions can imagine the moment when the first spaceship from another planet comes into earth orbit and makes contact with the radar screens of a waiting world. A prodigious, prophetic literature of space fiction – from Jules Verne and H. G. Wells to Kurd Lasswitz and Sakyo Komatsu – has accustomed the imagination to the feasibility of interplanetary journeys; and the visual images of films and television have extended this imaginative capacity by establishing the basic stereotypes of universally popular fantasies about the future. In fact, the gap between the possible and the probable is at times very small – at least for the world audience that follows the heroic adventures of Captain Kirk in the television episodes of *Startrek*. The never-failing powers of the sky-god and the never-ending successes of the U.S.S. *Enterprise*, far out in deep space and many light years away from Starfleet Command, combine all the attractions of the marvellous with the comforting assurance that the human race will go forward for ever and ever from new horizon to new horizon.

The belief that there are new horizons in time has been central to the idea of the future from the earliest theories of human progress in the eighteenth century to the most recent forecasts of the futurologists; and it has been at the centre of the tale of the future from the European interest in Mercier's *L'An 2440* in the 1770s to the world audience for the *Startrek* adventures in the 1970s. The evolution of the idea of the future is a complex story of arrivals and survivals, of the inventing of new modes of writing about the future, and the regular entrance of new themes. Any account of the ways in which the pattern of expectation has been formed and reformed during the last 200 years must, therefore, give special attention to those specific differences in origins, intentions, methods and themes that produced such distinctive styles as the tale of the war-to-come, the utopian prophecy of the world state, and the present-day forecasts of the futurologists. Hence, the account of these fictions and predictions must explain the quite separate histories of these various ways of looking at the future.

This book surveys the three main stages in the course of futuristic literature. The first stage opens with the ideas and the literary innovations that lead to the establishment of the tale of the future in the 1830s; it goes on to the beginnings of modern science fiction with the appearance of Jules Verne in the 1860s, and it ends with the supreme achievements of H. G. Wells about the end of the century. In parallel with this growth in popular fiction there is the very different course of development in utopian fiction; and the second part of this book has, therefore, to examine the political and social influences that produced the classic ideal states of the future in the nineteenth century. It goes on to consider the entirely different conditions that brought about the first essays in forecasting in the 1870s and led to another of Wells's extraordinary successes in *Anticipations* in 1901 – the first popular exercise in prediction in the history of futurology. The rest of the story is the continuing evolution of futuristic literature throughout the twentieth century. The third part of this book will, therefore, pay special attention to the major changes since the end of the First World War. These are: the sudden ending of the old-style tales of future warfare and the almost total disappearance of the progressive ideal states in the 1920s, the ominous prophecies of life in most undesirable brave new worlds of the future in the 1930s, the increasing popularity of science fiction since the end of the Second World War, and the rapid development of futurology in the last two decades.

Finally, there are the many unprecedented changes and developments of the last 30 years that have had far-reaching effects upon futuristic literature. When Neil Armstrong set foot on the Moon, it was evident that the facts of science had overtaken the old fantasies of the space adventure story. Today, when the futurologists give their estimates of economic growth or of the increase in world populations during the last decades of the twentieth century, it is equally evident that the idea of the future has reached the stage when prediction takes over from fiction. At this point the history of all our tomorrows comes to a temporary end. What began with the literary prophecies of the 1770s ends for the moment with the professional predictions of the 1970s. *Felix faustumque sit!*

PART ONE

Genesis and Evolution

I

The Discovery of the Future

Although the modern style in futuristic fiction begins with the European success of *L'An 2440* in the 1770s, there were prophets before Sebastien Mercier. The hesitant, clumsy handling of their anticipations demonstrates the problems of using the device of imaginative projection before the idea of the future had become a familiar element in European thinking. These preliminary exercises in the genre have their analogue in the state of the drama before the building of the theatres in London; for all these early stories suffered for want of the solid support and direction that a well-established literary convention can give.

The first appeared in the May of 1644: *Aulicus his Dream, of the Kings Sudden Comming to London.* The author, Francis Cheynell, was unable to foresee how he would one day so disturb the monumental urbanity of the *Dictionary of National Biography* that his record could only describe him as a fanatic, a rabid Puritan, noted for his zeal and bitter temper. Cheynell wrote what seems to be the first propaganda piece in the history of futuristic fiction: a six-page tract in the form of a dream, a nightmare vision of King Charles triumphant in London and the menace of his Irish regiments waiting outside the city.

The first report from the future, then, was a rudimentary affair which vanished into obscurity; and the same judgment can serve for the second account of time-to-come, *Épigone, histoire du siècle futur*, a two-volume publication of 1659 from another forgotten writer, Jacques Guttin. He is in a sense the first historian of the future, since he places his tale in an era when the Empire of the Clodovistes stretches from the Caspian Sea to the North Pole and from the African deserts to the east coast of America; but his vast panorama of a future France is no more than background for a tale of romantic love and knightly adventure, 636 pages in all, and it was never finished. Length seemed to matter at this stage in the course of futuristic fiction, since the next anticipation was planned as a six-volume enterprise, although only the 525 pages of the first volume ever came off the press, in 1733. The world is much indebted to some person unknown for the merciful act of suppression that terminated Samuel Madden's *Memoirs of the Twentieth Century, being Original Letters of State under George the Sixth*, since the tedious, wordy and boringly repetitious narrative presents a dreary vision of the future in which a Protestant bigot lets malice have its way. So, France is in a desperate condition – a weak King, predatory nobles and a starving populace. In Italy the Pope rules as a despot; the Jesuits are powerful throughout Europe and supremely powerful in Russia; and everywhere the power and prestige of the United Kingdom command respect and regard. One forecast has a slight interest. The improved telescopes reveal the world of the Moon as clear as can be:

> Not only the Hills, Rivers, Vallies, and forests, but real Cities in the Moon, that seem nearly to resemble our own and what is still more, even Mountains and Seas in Venus and the other planets. Nay some of our Astronomers have gone so far, as to aver, they could distinguish the Times of Plowing, and Harvest there, by the Colour of the Face of the Earth, and to specifie those Times, that others might make a Judgement of their Observation.[1]

Thirty years later, in 1763, that whisper of change grew a trifle louder in the anonymous account of *The Reign of George VI, 1900–1925*. This book, which marks the end of the palaeolithic period in futuristic fiction, is a delightful but neglected story. It is filled with the most attractive and unwitting anachronisms, a true literary fossil from the last evolutionary epoch before the appearance of the developed form. And yet this imagined history of a future British monarch shows, already well established in the embryonic form, the essential structure of the political prophecy; for ideal states of the future, like the predictions of the next great war and descriptions of coming catastrophes, are of their nature exemplary. The Tory heaven-to-be of this unknown author demonstrates the mechanics of the device. Here, as in Sebastien Mercier's *L'An 2440* or in Edward Bellamy's *Looking Backward*, the historical account presents the united and contented society that has emerged from the obscurities of the future, so the propaganda suggests, because the truths of political philosophy (as an author understands them) or the forces of history (as an author defines them) have established a new order in the world.

Again, the events in *The Reign of George VI* reveal that fusion of narrative method with political theory that joins body with soul in the exemplary vision of the future. In projections of this kind–both utopian and dystopian–some dominant theory of one form or another defines the contours of the coming society and some favoured contemporary style of narrative decides the telling of the story. For instance, the cinematic techniques of cutting and flash-back dominate both Huxley's *Brave New World* and Orwell's *Nineteen Eighty-Four*, and just as Huxley started from the Wellsian image of the superstate, so Orwell drew some of his most effective ideas from James Burnham on the managerial revolution. It was the same with the design for *The Reign of George VI*–a working model for demonstrating the political theories of Bolingbroke in the *Idea of a Patriot King*, the whole written in the spirit and the style of the first popular *History of England* which David Hume had been turning out at intervals down to his sixth volume in 1762.

The Hanoverian harmonies of *The Reign of George VI* contain an instructive contrast. The major innovation of the future-perfect theme goes with a defensive attitude to the narrative which looks back to literary tradition for support. The author is so unsure of his invention that he clearly feels it necessary to discuss his intentions and methods. He begins with self-conscious explanation: 'With regard to the tendency of the following history, as it is taken up at a what's-to-come period, and begun at an aera that will not begin these hundred years, it may be necessary to say a few words, whether critical or explanatory, whimsical or elaborate, shall be entirely submitted to the determination of the reader.' These are the hesitations of a prophet before the event. Although the author had found a satisfactory prescription for political progress in the ideas of Bolingbroke, the oracle spoke with an uncertain voice. He lacked that other necessary power – the certainty of material change which validates the revelation of the future, since it agrees with the general expectations of the time and so makes more universal, more probable and acceptable what in the political theory must be both partial and particular. For want of that inner light the Tory propagandist had to persuade his readers to accept his extensive reorganization of time-to-come; and he laboured through some 1100 words in the preface to show that he was following in the approved tradition of *Gulliver's Travels*, suggesting that 'in the course of the following sheets, the reader's own reflection must frequently assist him in the elucidation of particular circumstances – for in performances of this nature it is totally impossible to be always as clear as a person could wish.'

With a nod and a hint the anonymous author begins his Georgiad of 1763 with a chapter on the course of history up to the end of the nineteenth century. The modern reader will find that little has changed – political jobbery and parliamentary factions continue into the twentieth century; the American colonists are the loyal subjects of the Crown; frigates and three-deckers control the seas and the monarchs of Europe lead their troops into battle. Nevertheless, the author made one correct guess. Going forward from the beginning of canal construction in 1761, he reports that his most prudent monarch has encouraged the building of waterways. In consequence, 'rivers that formerly were almost useless were now navigated by large barges, which increased the trade of innumerable towns, and raised in many places new ones. The canals which were cut joined rivers, and formed a communication between every part of the kingdom. Villages grew into towns, and towns became cities.' And yet that fact of industrial and commercial growth made for no more than a modest increase in national wealth, and the spreading urbanization did not suggest any marked rise in population. Did the author live on to read of

Jenner's vaccine? Did he read Malthus on population? What did he make of the great transformations he may have seen – revolutions in America and in France, extraordinary developments in technology and medicine, new ideas in poetry and prose – all those many changes that diverted the course of history from the tranquil, static future world of an unknown Tory propagandist?

If the author lived to see the consolidation of the new literature of the future throughout Europe at the turn of the century, then he must have realized that *The Reign of George VI* had demonstrated the first application of political theory to the practice of the new genre; for the guiding principles in his prophecy derived from the *Idea of a Patriot King* which the Viscount Bolingbroke, formerly Henry St John, had published in 1749 for the instruction of the Prince of Wales. His initial proposition argued for the superiority of monarchy to all other forms of government and for the perfection of constitutional monarchy. It followed, therefore, that a wise, able and dedicated monarch would play King David to a regenerated nation, and that the Patriot King would do everything necessary for the well-being of Church and State. This utopian theorizing found faithful realization in the character and actions of the imaginary George VI. Who, then, is the Patriot King that every monarch should wish to be? 'A Patriot King', Bolingbroke wrote, 'is the most powerful of all reformers ... the sure effects of his appearance will be admiration and love in every honest breast, confusion and terror to every guilty conscience, but submission and resignation in all. A new people will seem to arise with a new king.'

In *The Reign of George VI* the scenario dissolves from the events of the nineteenth century to reveal the alarm of the United Kingdom when a powerful Russian invasion force lands on the coast near Durham. The new King descends upon a corrupt and terrified Parliament, drives the Speaker from the chair and speaks his mind forthwith to the House: 'I shall place myself at the head of my troops, and act for the honour and good of my country. But let those traitors, that dare form machinations against the public peace, dread the indignation of an injured and enraged Sovereign.' The King storms out, reorganizes his forces and takes control of the Bank of England. From the North the news comes that Count Schmettau, the Russian commander, has ordered the sack of Durham. 'The Russians broke into all the houses, and were guilty of every excess. Their cruelties were unheard of and unparalleled; the most tender age was no defence against these merciless monsters; old men, women, and children were butchered in cold blood, in the most shocking manner.'

Thereupon the King marched north and on December 23rd, 1900, engaged the enemy at Wetherby. On that day and for the first time he showed the decisiveness and generalship that were to make him the greatest captain of the twentieth century. His courage had a conclusive effect on the battle: 'Never man performed greater feats of personal valour; he had three horses shot under him, and as he was going to mount a fourth was near being shot by a Russian grenadier, but his carbine missing fire the King shot him dead.' After this the enemy surrenders; and the King at once rides south to be admiral to his fleet. He boards the *Britannia*, 'which was without exception the finest ship in the world; she carried 120 brass guns, and, in the opinion of the best judges, was so well built and manned that no single ship could live near her.' And, once again, the King is triumphant. In a three-hour engagement he crushes the Russian fleet:

> Nine of their line of battle ships were taken, three sunk, and two burnt; forty transports were also taken, and several sunk. Thus did this young and gallant Monarch, with all the courage, conduct and skill of an experienced Admiral, defeat the enemy's fleet, which was so much superior to his own. This second victory raised the fame of the King to the highest pitch, changed the face of affairs, and spread a general joy through the breasts of all his subjects.[2]

Chapter by chapter this wish-fulfilment fantasy rolls on, every event a demonstration of the Bolingbroke idea. The Patriot King is the true father of his people; hospitals, colleges and new towns arise throughout the land; the arts and literature flourish in keeping with the canons of eighteenth-century taste; trade and manufactures multiply. Abroad, he extends the benefits of

British rule, as more and more of the European countries submit to the Hanoverian hero. France and Spain are conquered, and by 1920, 'all Europe trembled at the name of George'. Finally, the Bolingbroke theory appears most forcefully in the moment of the French surrender, when George VI shows his new subjects that he is no despot. In proof of the Bolingbroke principle that the supreme good depends on the readiness of the sovereign to limit his powers, the French are seen to enjoy all the liberties of the British people. The fortunate French receive the benefits of trial by jury, an uncensored press, and a well-paid, independent judiciary. As the French and most of Europe settle into the new life of the *pax britannica*, the history concludes with a note for posterity: on all sides peace and happiness reigns; and the measure of the King's achievements is in 'the universal praise that is bestowed on his memory by all foreign historians. His name was as dear to France as it was to Great Britain. Fortunate nations to possess a king formed by nature to make the world he governed happy!' *L'état c'est lui.*

A light shines in the darkness. The saving power of right principle brings social order out of universal chaos. It creates a future that will be for ever British or French, Capitalist or Socialist, Protestant or Catholic, as every author of every ideal state has the right to decide. There, far away in the future, the perfect commonwealth exists for all the loyal subjects of a British monarch, or for the citizens of the true republic, or for all mankind united in the federation of the world. Whatever form the imagined future may take, the shaping influence is the power of myth. These visions of the secular paradise repeat the *in principio* of sacred history within the context of civil society: they recount the creation of the world in the image of political theory; they renew the cycle of heaven and hell, sin and redemption, in an historical narrative that begins with the original error of a disobedient people or of an unjust society; and they close with the saving grace of an ideal system that restores the natural harmony between citizen and society. So, the events in *The Reign of George VI* start from the primordial moment of crisis and chaos when 'all eyes were turned on the King, as the only pilot in so terrible a disaster. It was impossible to be guided by a better; and had not Britain possessed a

Sovereign of such singular intrepidity and prudence, she would have seen her last days.'

In a similar way, but in another continent and in a technological society, the better future of Edward Bellamy's *Looking Backward* begins at that moment in time-to-come when

> ... at last, strangely late in the world's history, the obvious fact was perceived that no business is so essentially the public business as the industry and commerce on which the people's livelihood depends, and that to entrust it to private persons to be managed for private profit, is a folly similar in kind, though vastly greater in magnitude, to that of surrendering the functions of political government to kings and nobles to be conducted for their personal glorification.

And so, in the sequence of an idealized history the aboriginal condition of society requires an appropriate conclusion in the vision of the ideal state that represents the best conceivable way of life.

These versions of the life-to-be are the cosmogonies of the new industrial societies.[3] They are both explicative and communicative – acts of mediation between the best self in any society and the far better state of things in time-to-come. Dearly beloved, the texts suggest, all desirable things will come to pass in the fullness of time as soon as the citizens – or the rulers or all human beings – begin to practise the precepts of right order and social justice. That message has been handed down in many forms during the last 200 years, from the first proposals in *The Reign of George VI* and *L'An 2440* to the most recent solutions from B. F. Skinner in *Walden Two* and from Huxley in *Island*; but no matter what the differences in social values and in general philosophizing may be, these anticipations seek to provide answers to the most intractable of human problems – how to refine the passions and eliminate the prejudices of mankind.

The critics were less than dispassionate in their treatment of the anonymous author of *The Reign of George VI*. According to the *Monthly Review*, it was 'probably the work of some callow writer, of more vivacity than judgement or genius'. And the *Critical Review* reported that the story failed, 'because the virtues, the achievements and the grandeur of this same George VI never can exist in one person'. The anonymous author had committed the major errors that Sebastien Mercier was careful to avoid in *L'An 2440*: he had limited his narrative to an indulgent vision of a successful British imperialism and a miracle-working monarch.

It is worthy of note that, although the earliest specimens of futuristic fiction were in English, the glory of discovering the true potentialities of the genre belongs to the French – to Sebastien Mercier who established the first satisfactory model of the new fiction; to Cousin de Grainville whose story of *Le Dernier Homme* in 1805 introduced the theme of The Last Man to European literature; and to Félix Bodin who first examined the course of futuristic fiction in *Le Roman de l'avenir* in 1834. This sudden flowering came naturally to French literature, since the great debate about nature and society, which had characterized France in the eighteenth century, posed questions that could only find their solution in a theoretical future. Hence, Sebastien Mercier could claim with some justification that in *L'An 2440* he had anticipated the French Revolution, since his better France of the twenty-fifth century demonstrated the political theories of the Enlightenment.

The rapid success of *L'An 2440* marks the first triumph of the man and the moment in the course of futuristic fiction; for Mercier presented his vision of universal felicity to the small, literate world of eighteenth-century society at a time when contemporary ideas about the progress of mankind were beginning to encourage more ambitious hopes of future change. The utopian prophecy of *L'An 2440*, the first influential story of the future in world literature, became one of the most widely read books in the last quarter of the eighteenth century. The translations into Dutch and German, the English and American editions, and the many imitations show how Mercier beguiled his readers in two continents with the vision of a dream come true. The entry of the dreamer into the perfect world of the twenty-fifth century had inaugurated a new phase in the long unwinding of European expectations. The detailed account of a world of peaceful nations, constitutional monarchs, universal education and technological advances was an extension of the scheme Bacon had presented in *The New Atlantis*. Mercier's belief in the continued progress of institutions and in the certain improvement of society derived from a most persuasive ideology of man and nature that had spread through Europe in the eighteenth century.

In the beginning, as Pope wrote in a familiar couplet, 'Nature and Nature's laws lay hid in night: / God said, *Let Newton be!* and all was light.' Pope meant that Newton's *Principia Mathematica* had explained the laws of motion in the universe and had shown that these can be verified by observation and experiment. By this act of revelation Newton became prime mover in a series of ideas that had great effect upon the way in which the philosophers and historians of the eighteenth century looked on their world and on the course of history. The grand design of the universe suggested that there might be some comparable scheme in the mechanics of society, and the entirely new principle of the rate of change in the physical world encouraged the assumption that some rationally explicable laws of social motion might be at work in human history. A succession of thinkers began to examine the conditions that made for the improvement of society; and the vocabulary of the educated acquired luminous phrases that spoke about the evidence of historical facts, the general system of the sciences and the arts, the constant and universal principles of human nature, the gradual and progressive augmentation of wealth, and the progressive improvement of mankind. 'Finally it is to be believed', Voltaire wrote in his 'Essay on the Manners and Mind of Nations', 'that reason and industry will always bring about new progress.'

From the evidence of history and the fact of technological advance there emerged a theory of progress. For some this was a progressive and libertarian movement that mankind could direct and influence; for others it was a progressive and deterministic process that would evolve in its own predetermined way. This immensely powerful idea, which was to become sacred doctrine in the nineteenth century, taught that mankind had advanced, is advancing and would advance – or should be encouraged to advance – still further. And in a singularly appropriate way the world heard the first major formulation of the new theory on December 11th, 1750, from a young man of twenty-three, a theological student at that time, who read a paper before the Sorbonne which he described as *A Philosophical*

Review of the Successive Advances of the Human Mind. The author was the celebrated Turgot, a founding father in the philosophy of progress, who ended his days as Minister of Marine and Controller-General of Finance.

Turgot addressed his audience of clerics with the total confidence of a man who had the year before become a Bachelor of Theology, and in the name of Christianity he insinuated a new doctrine of human perfectibility: 'The whole human race, through alternate periods of rest and unrest, of weal and woe, goes on advancing, although at a slow pace, towards greater perfection.' Wherever he chose to look in history, Turgot found abundant evidence to prove his claim that all civilized communities had struggled upwards from barbarism into the full light of civilization. The Phoenicians, as he showed, were an example of the relationship between enterprise and environment, because they had escaped from the limitations of a barren coast. Their evolution repeated itself in modern times; for 'colonies are like fruits which cling to the tree only until they have reached maturity: once they had become self-sufficient, they did what Carthage was to do later, and what America will one day do.'

The confident manner is not a trick of style. Turgot writes with all the conviction of the philosopher who knows that he has discovered the secret of the universe; and he is so certain of the causes at work in human affairs that he forecasts the disruption destined to take place between the United Kingdom and the loyal colonies across the Atlantic. As the precursor of a long line of millenary thinkers – Condorcet, Priestley, Godwin, Fourier, Saint-Simon, Comte, Enfantin – Turgot handed down the idea of progress as an all-pervasive and all-powerful motive force in history. As the Newton of a new social philosophy, he concentrated the entirety of human achievement into his unique law of socio-dynamics, the 'mutual dependence of all truths' which promotes the advance of the human race. And like a latter-day Aquinas in the halls of the Sorbonne he ended his discourse with a profession of faith: 'Open your eyes and see! Century of Louis the Great, may your light beautify the precious reign of his successor! May it last for ever, may it extend over the whole world! May men continually make steps along the road of truth! Rather still, may they continually become better and happier!'[4]

Those hopes for posterity – better, happier, more secure and more prosperous – were reflected in the tale of the future as it developed towards the end of the eighteenth century. The first major statement of the new optimism appeared in *L'An 2440*; and there Mercier showed that he had listened to the eternal rhythms of Nature and Reason. Out of the ideas of Turgot, Montesquieu, Voltaire and Rousseau he had composed his symphony of universal peace and progress. His repeated references to rationality and humanity reflect the hopes of many Frenchmen in the years before the guillotine and the Terror: order, decency, affability; venerable words, sublime truths, enlightened understanding; delightful sentiments of virtue and humanity, laws dictated by reason and humanity, the influence of knowledge and progress. These are the sacred words and the authentic accents of eighteenth-century optimism. 'But oh! heavens! what do I perceive! How beautiful the banks of the Seine! My enchanted vision is gratified with the finest monuments of art.' And with this cry of delight the narrator moves on from marvel to marvel, exclaiming at the happiness and justice and general wisdom of an ideal France.

The anonymous narrator is the first true time-traveller in literature. He exhibits, already well developed, the device of the dreamer which became a standard means of introducing many later tales of anticipation. His story has the added distinction that it was the first ideal state of the future to be banned by the censor. In 1778 the King of Spain issued a *Real Cédula* forbidding 'the introduction and sale of a book entitled *The Year 2440*' on the grounds that it was blasphemous and anarchic in tendency:

> The idea of this impious writer is to imagine a dream, and after it he awakes in Paris in the year two thousand four hundred and forty; and by means of this invention he describes the supposed state in that time of the Court of Paris, the Monarchy of France, Europe and America – affecting disillusionment and supposing changes to have taken place in all ecclesiastical, civil and political government.

The summary was exact. Mercier began his report from the twenty-fifth century with the necessary circumstantial evidence: it was midnight and the narrator falls asleep tired out; he sleeps profoundly, and he awakes an age after with trembling hands and tottering legs; his brow is furrowed and his hair is white. No suspended animation for him. He staggers out, an old man going to explore a new world, and he sees from the engraving on a monument that he has arrived in *The year of Our Lord M.MIV.CXL.*

The eleven editions of *L'An 2440* between 1771 and 1793, the four English translations, the two American editions, the Dutch and German translations, the many imitations – all these show that the Mercier scheme coincided with contemporary ideas about the most desirable social system. The opening scenes reveal an austere and simple way of life. There are sumptuary laws against ostentatious living, clothes are no longer extravagant, and simple hairstyles have replaced the wig. Education is the concern of the state; Latin and Greek have vanished from the schools, their place taken by the international languages – English, German, Italian, Spanish. The citizens are all deists in religion; they are pious, careful of the sick and the old, devoted to the service of the Supreme Being, the Eternal Ruler of the World; and they make voluntary contributions to the national exchequer. This perfect social scheme is part of a greater world order, for peace and amity prevail throughout the planet from China to Peru. Thus, the peoples of the Americas have recovered their ancestral lands, and a Montezuma reigns once more in Mexico City. There are two kingdoms in Russia, Poland is a strong monarchy, and Portugal has become part of the United Kingdom. All the injustices of the past have gone: arbitrary government, imprisonment without trial, religious persecution, distinctions between classes. The rulers of the nations are constitutional monarchs – wise, just, dedicated to the service of their grateful peoples:

> Sovereigns have, at last, been prevailed on to listen to the voice of philosophy; they are now linked together by the strongest ties, they are become acquainted with their own interests. After so many centuries of error reason has at last

resumed her seat in their souls. They have opened their eyes to those duties which the safety and tranquillity of their people exacted from them. They have made their glory consist in governing wisely, wishing rather to cause the happiness of a small number than to gratify the frenzied ambition of ruling over desolated countries filled with aching hearts, who always detested the usurped power of the conqueror. Kings have with one accord set limits to their empire: those limits which nature herself seemed to have assigned by separating every respective kingdom by seas, forests, or mountains. They now understand that the less extensive the territory, the more it is likely to be wisely governed.[5]

Long before H. G. Wells argued in *A Modern Utopia* that 'synthesis is the trend of the world', Sebastien Mercier had foreseen the day when the resources of technology would ease the task of reason and virtue by linking all mankind in one fraternal union. For the first time in the history of utopias a writer had assumed that the combined logic of humanity and of science would inevitably lead to concord and co-operation throughout the planet. Mercier had broken out of the self-contained and self-sufficient utopia of city-states and faraway islands. The powerful appeal behind his vision of universal brotherhood was the assurance that this world union was a possible achievement within the reach of all men of good will: 'We all view each other as brethren and friends; both the Indian and the Chinese are our fellow-citizens as soon as they tread on our soil. We accustom our children to look on the universe as one family, sheltered under the protection of one common father ... Men have learned to love and esteem each other.' The hopes belong to the eighteenth century, but the means of achievement lie with the technologies of the future. There are regular maritime communications between all the nations of the world, and the engineers of the new age have 'effected the long meditated project of forming canals from the Nile to the Arabian Gulf, without any cause to fear the overflowing of the gulf by that communication. By this means Egypt is open to every nation in the world, and is become the great mart of commerce between Europe, India and Africa.' But the hopes were more exact than the forecasts. Mercier looked ahead, for example, to the time when a new canal system would link the North Sea to the Mediterranean; and yet the sole advantage he could imagine for this great enterprise is that 'by this means the treasures of commerce are carried from Amsterdam to Nantes, and from Rouen to Marseilles'. Again, desiring to obtain the immense benefits of mechanical power, Mercier reports that in his future world there are 'all sorts of machines for the relief of man in laborious works, and capable of much more force than those in our times'. This is still in the tradition of *The New Atlantis*. Mercier makes an act of faith in the powers of science, but fails to discern the true shape of the hoped-for improvements. The failure was inevitable, because Mercier had designed his ideal state twelve years before the spectacular successes of the first balloon flights taught the would-be prophets the new methods of divination.

The story of that primal period in the history of modern technology reads like a fairy tale; for once upon a time and for most of human history mankind had dreamed of the conquest of the air, and then quite suddenly the dream came true during the summer and autumn of 1783, when the inhabitants of many European cities saw the first balloons drifting silently through the skies. Like the discovery of America, the balloon ascents were the promise of still greater advances. It was a moment in world history when the imagination began to move freely in the new dimension of time-to-come; and Sebastien Mercier was there to report on what took place. For the space of six months Paris was the Cape Canaveral of the eighteenth century as reports spread everywhere of the new Montgolfier balloons; and then on November 21st came the astounding news that two men, Pilâtre de Rozier and the Marquis d'Arlandes, had taken off into space. Sebastien Mercier gave this report of the ascent:

> A memorable date. On this day, before the eyes of an enormous gathering two men rose in the air. So great was the crowd that the Tuileries Gardens were full as they could hold; there were men climbing over the railings; the gates were forced. This swarm of people was in itself an incomparable sight, so varied was it, so vast and so changing. Two hundred thousand men, lifting their hands in wonder,

admiring, glad, astonished; some in tears for fear the intrepid physicists should come to harm, some on their knees overcome with emotion, but all following the aeronauts in spirit, while these latter, unmoved, saluted, dipping their flags above our heads. What with the novelty, the dignity of the experiment, the unclouded sky, welcoming as it were the travellers to his own element, the attitude of the two men sailing into the blue, while below their fellow-citizens prayed and feared for their safety, and lastly the balloon itself, superb in the sunlight, whirling aloft like a planet or the chariot of some weather-god – it was a moment which never can be repeated, the most astounding achievement the science of physics has yet given to the world.[6]

The sudden achievement of balloon flight had provided indisputable evidence of the power of science and – even more important – had penetrated to the most profound depths of the human psyche. The sky, for millennia the abode of gods and the sacred place to which only the elect could ascend, had yielded to human inventiveness. For Diderot the balloon ascents of 1783 were a promise that men would one day go to the moon. For Louis XVI they were sufficient cause for Letters Patent of Nobility:

> Louis, by the grace of God, King of France and of Navarre, to all present and to come, greeting: The aerostatic machines invented by the two brothers, the Sires Étienne-Jacques and Joseph-Michel Montgolfier, have become so celebrated ... have had such success that we have no doubt but that this invention will cause a memorable epoch in physical history; we hope, also, that it will furnish new means to increase the power of man, or at least to extend his knowledge.

The *Encyclopaedia Britannica* echoed the royal hopes in the article on 'Air Balloons' which James Tytler, the first British aeronaut, wrote for the Appendix of 1784: 'By this invention the schemes of transporting people through the atmosphere, formerly thought chimerical, are realized; and it is impossible to say how far the art of navigation may be improved, or with what advantages it may be attended.' About the same time, when news of the Montgolfier achievement reached Thomas Jefferson, he wrote to his cousin in Virginia to suggest some of the likely uses for the new balloons: 'Traversing deserts, countries possessed by an enemy, or ravaged by infectious disorders, pathless and inaccessible mountains; the discovery of the pole which is but one day's journey in a balloon, from where the ice has hitherto stopped adventurers.'[7] In England the begetter of Gothic fiction, Horace Walpole, reported to Sir Horace Mann that 'Balloons occupy senators, philosophers, ladies, everybody. France gave us the *ton*.'[8] And an excited French poet wrote: 'Do not talk about impossibilities, for nothing is impossible to determined effort ...' and in a lamentable couplet he summed up the wonders of the day:

> Cook marche aux fonds des mers. Montgolfier vole aux Cieux;
> Ouvrez-moi les Enfers, j'éteindrai les feux!

In Germany the writer Christopher Wieland coined the word *Aeropetomanie* to describe the general feeling of excitement and wonder at the new balloons. 'The marvels of our century', he wrote, 'seem to press ever more closely on one another, and the nearer we come to the end of the century, they seem to become ever greater and ever more wonderful.'[9]

The early balloon ascents mark the first movement of expansion in the new literature of anticipation. After 1783 the tales of the future multiply rapidly and there is an equally swift development in the predictions of coming advances in technology, especially in engineering. From the start the fantasies and the forecasts go forward in their parallel courses, as more and more writers describe the world they would like to see, or seek to analyse the factors that will decide the future state of society, or examine the trends in population that will affect all human beings on the planet.

This new sense of expectation is apparent in the first account of the Montgolfier balloons to be published in the United Kingdom. This appeared at the end of 1783 – *The Air balloon, or a treatise on the Aerostatic globe lately invented by the celebrated Mons. Montgolfier of Paris* – and the author, William Cooke, begins by stating that the balloons were

> ... an experiment which in a very few ages back would have filled the world with amazement and wonder, and perhaps have sent the inventor to his grave with ignominy and disgrace. The times, however, in this respect, are more enlightened; for whilst this phaenomenon produces novelty, and opens a wide field for speculation and improvement – it gives due honours and rewards to the Philosopher.

There were similar expectations in the first balloon adventure story in English, *The Aerostatic Spy, or Excursions with an Air Balloon* which appeared in 1785, its anonymous author declaring: 'The invention of Aerostatic Machines is of such a nature as to have formed a new Aera in the History of Science. The utility of this Invention, yet in its infancy, remains to be exhibited by subsequent Improvements; and that Improvements will take place there can be little reason to doubt.' That proposition met with general agreement in France, Germany, the United Kingdom and the United States. That is, wherever the scientific base of a nation was extensive and wherever there were close links between experimental science and technological application, then the belief in progress was strong enough to inform, to encourage and to support the *Zukunftsroman*, the *roman de l'avenir* and the *tale of futurity*, as the new futuristic fiction was called.

Wherever writers described the shape of things-to-come or wrote about the new inventions, there was an accepted international vocabulary that spoke of change, invention, improvement and a new era for mankind. That was, in substance, the main point which the Marquis de Condorcet put to the Europe of 1784 in his *Fragment sur l'Atlantide*, for this most dedicated of Baconians maintained that the world was close to realizing the great design set out in *The New Atlantis*. In fact, thanks to the rapid advances in knowledge, there was good reason for expecting that the new order would begin in his own time; and looking ahead to future times, he argued that one advance would lead to another. 'Let us suppose,' he wrote, 'that Euler and his abilities were to be transposed to some future period, then methods of which we know nothing today and an immense body of truths not yet discovered would enable this future being to start on the solution of problems that we cannot even conceive of proposing to ourselves.'[10]

Condorcet was right. The Montgolfier balloons and the steam-engine, then the power loom and lithography, then Jenner's vaccine and all the other discoveries of that time were ample evidence that human society had passed a point of no-return and was at last moving into a new kind of existence. The first manifestation of this came in a flood of poems, plays, stories, and prints on the popular subject of balloon travel. In 1784 the Paris theatres were quick to amuse the public with plays about the adventures and romantic experiences of the new aeronauts in such comedies as *Le Ballon, ou la Physico-manie* and *Le Siècle des ballons*; and the Germans, who usually followed the French in these matters, turned out similar plays – in 1786 Christian Brezner produced *Die Luftballe* for the citizens of Leipzig, and in that same year Max Blumhofer wrote the first space operetta, *Die Luftschiffer*, for the Imperial Court Theatre of St Petersburg. As the redoubtable empress Catherine II listened to the chorus singing the marvels of space travel, she may have reflected on the enthusiasm for future change then sweeping through Europe. She may even have agreed with the popular belief that the anonymous author of *The Aerostatic Spy* repeated at the beginning of his imaginary journey round the world: 'Man was created to render the Elements subservient to his Use and Convenience. Earth, Water, and Fire for Ages owned his Sway; but it was reserved to the Men of the present Times to add the Element of Air to their Empire, and soar sublime in fields of trackless AETHER.'

Ever since the first balloon ascents of 1783 and the first wave of balloon stories the course of futuristic fiction has been the record of an ever closer and ever more complicated relationship between science and society. Nowadays, when our frequent anxieties about the consequences of technological inventiveness cause many to fear the future, it is difficult to enter into the far freer and far more hopeful spirit of the time when the new sciences seemed to promise the most desirable improvements in the condition of mankind. In those days the balloon and the steam-engine were the signs of what would come; and evidence for this can be seen in two events of 1786 – the new chapter on balloons which Mercier wrote for the second edition of *L'An 2440* and the visit of James Watt and Matthew Boulton to Paris. They had come at the request of the French government to discuss the construction of steam-engines in France and, in particular, to suggest improvements for the hydraulic pumping machinery at Marly on the Seine. They met Lavoisier, Laplace and Gaspard Monge; and although there is no detailed report of that encounter between practice and theory, there is every reason to assume that the English manufacturer and the Scottish engineer (like the chemist, the astronomer, and the mathematician) must have looked on the massive structure at Marly as the wonder of an age that was coming to an end.

In the palaeotechnic period, before Watt invented the separate condenser, the most powerful prime mover ever constructed was the Great Machine of Marly, the hydraulic system which Louis XIV had had built in 1682 to provide water for his fountains at Versailles. This marvel of misapplied engineering skill generated about 75 horsepower and, before the wooden parts began to wear, it could raise one million gallons of water a day to a height of 502 feet. By 1786, however, those who understood the recent advances in technology could perceive the world of the future in the whirling planet wheels of Watt's new beam-engines. The old order was vanishing in the smoke and steam of the industrial revolution, as the new breed of mechanical engineers set to work to develop the steam-engine. Already in 1786, across the Atlantic, the legislature of Pennsylvania had received a petition from the Delaware wheelwright, Oliver Evans, for exclusive rights to apply his improvements in the steam-engine to flour milling and road transport. In the following year Maryland gave Evans authority to develop his plan for constructing steam-wagons, and the inventor promised 'that the time will come when carriages propelled by steam will be in general use, as well for the transportation of passengers as goods, travelling at the rate of 15 miles an hour, or 300 miles a-day on good turnpike roads.'[11] The tales of the future were soon to make good that promise.

2

To the Last Syllable of Recorded Time

Two centuries ago, when the first utopias of the future were appearing, the message from the laboratory and the workshop was that every advance in science would lead to a better life for all. And so, it was goodbye to Plato and goodbye to More, as the designers of ideal states of the future looked ahead to the great changes that were sure to come. The first to convey the revelation of the new sciences was Sebastien Mercier. In the revised edition of *L'An 2440*, which he prepared in 1786, he included a new chapter on *L'Aérostat* in order to show how the Montgolfier balloons would one day change world communications.

The chapter opens with the arrival of 'an immense machine which came on in full sail'. It was the flying machine from Peking, and when it had come down to earth, 'eight mandarins emerged from the cabin which was suspended beneath the aerostat'. Mercier then goes on in his usual way to enlarge upon the astounding discovery, the noble conquest that man has made of the third element, the slow but sure march of science: 'These *bird-men*, for that is the name they give to the aeronauts, have made themselves the inhabitants of the clear skies and the pure light; they travel through the storm, moving from one climate to another in twenty-four hours, traversing distances that used to separate the most distant countries.'[1]

This was the very April of European hopefulness, the Mercier moment which inaugurated the great outpouring of prophecies and predictions that were then taking on the new task of describing the shape of coming things. And in most of them science was the pace-maker and the scientist was the wonder-worker of the new age:

> From every shape that varying matter gives,
> That rests or ripens, vegetates or lives,
> His chymic powers new combinations plan,
> Yield new creations, finer forms to man,
> High springs of health for mind and body trace,
> Add force and beauty to the joyous race,
> Arm with new engines his adventurous hand,
> Stretch o'er these elements his wide command,
> Lay the proud storms submissive to his feet,
> Change, temper, tame all subterranean heat,
> Probe labouring earth and drag from her dark side
> The mute volcano, ere its force be tried;
> Walk under ocean, ride the buoyant air,
> Brew the soft shower, the labor'd land repair,
> A fruitful soil o'er sandy deserts spread,
> And clothe with culture every mountain's head.[2]

This programme for the future came from the United States, from Joel Barlow, one of the more liberal thinkers of that time, a prominent statesman who ended his days as American ambassador to France. He was a friend of Thomas Paine; and when that lifelong apostle of liberty was imprisoned by Robespierre, Barlow took charge of the manuscript of his *The Age of Reason* and arranged for it to be published. Barlow was then engaged on the great work of his life, an epic poem in ten books which he intended to be the most complete statement of all the latest ideas about the history and progress of mankind. Naturally, the description of the future in the tenth book of *The Columbiad* repeated the doctrines of Rousseau, Condorcet, Mercier and Thomas Paine. In one of the closing passages, which anticipated some of Tennyson's later hopes in 'Locksley Hall', Barlow looked ahead to the future epoch of reason and virtue:

> Yet there thou seest the same progressive plan
> That draws for mutual succour man to man,
> From twain to tribe, from tribe to realm dilates,

In federal union groups a hundred states,
Thro' all their turns with gradual scale ascends,
Their powers, their passions and their interest blends;
While growing arts their social virtues spread,
Enlarge their compacts and unlock their trade;
Till each remotest clan, by commerce join'd,
Links in the chain that binds all human kind,
Their bloody banners sink in darkness furl'd,
And one white flag of peace triumphant walks the world.[3]

This certainty of universal concord, the hopes for a renewed society of the future and for the benefits of technological improvement were the principal elements in a dynamic system of progressive thought then developing on both sides of the Atlantic; and together they provided both the motive power and the direction for the earliest futuristic utopias to appear after the first edition of *L'An 2440* in 1771. In Danish, Dutch, French, and German, one after another they promised that reason and science would lead mankind to a happier existence; and many of them, the German writers in particular, took the narrative technique of Mercier's dream as their model. These German versions of the *Zukunftsroman* made up the largest group of futuristic utopias that appeared in any country during the last decades of the eighteenth century: *Das Jahr 1850* (1777), *Die Frauenzimmer im neunzehnten Jahrhundert* (1781), *Beylage zu dem Jahre 2440* (1781), *Das Jahr 2440 zum zweitenmal geträumt* (1783), *Das Jahr 2500* (1794), *Die schwarzen Brüder* (1795). Outside Germany the wave of imitations rolled on through Denmark and the Netherlands – *Holland in't Jaar MM,CCCC,XL* (1777), *Anno 7603* (1785), *Het toekomend Jaar drie duizend* (1792).

These imitations of Mercier have only an antiquarian interest for the modern reader. They now repose in the rare book collections of national libraries, the silent evidence of a new direction in European literature. In fact, the hope for growth and originality lay with French and English fiction. In the United Kingdom the evolution of the new form had been very different, generally pragmatic in attitude and often topical in the choice of themes. After the initial philosophizing in *The Reign of George VI* the ideal state of the future vanished from the English language, not to reappear until the 1820s. Instead, there was a succession of projections in which writers tried out different effects from the Gothic and romantic to the immediate and political.

The first of the new stories came from an anonymous author who followed *The Reign of George VI* in 1769 with the first satirical forecast of the future, *Private Letters from an American in England*. It is the earliest example of the visitor from another time who finds that everything has worked out for the worst – Scottish immigrants, corrupt bishops and fanatical Methodists have ruined England. This anonymous author wrote in the favoured epistolary style of the day and his successors, who could have learnt something from *L'An 2440*, wrote as if Mercier had never existed. The device of the dreamer never appears in the new stories, and many of them have little, sometimes nothing at all, to do with the idea of technological progress. Again, most of them were entirely domestic in their subject matter and often didactic in manner, early examples of the intense communal activity that has always characterized great areas of futuristic fiction.

The first application of contemporary politics in futuristic fiction appeared in a pamphlet of 1778, *Anticipation: containing the Substance of HIS M-----Y's Most Gracious Speech to both H----S of P--L-----T, on the Opening of the approaching Session.* In the imaginary debate that followed after the members had returned from the Lords the would-be scourge of the American rebels, Lord North, urges the Commons 'not to forget that America is still the offspring of Great Britain: that when she returns to her duty, she will be received with open arms and all her faults be buried in oblivion'.

The first essay in the ironic view of the future came from Sir Herbert Croft who had the then original idea of using the graveyards of the twentieth century in order to comment on his contemporaries. The method was simplicity itself: in *The Abbey of Kilkhampton* of 1780 and in *The Wreck of Westminster Abbey* of 1788 the author presented 'a selection from the monumental records of the most conspicuous personages who flourished towards the latter end of the eighteenth century'. These first epitaphs from posterity clearly pleased the readers of the time. There were nine editions of the first book by 1788 and the second seems to have gone through three editions before 1790. No doubt it appealed to the eighteenth-century sense of the ironic to read an anticipatory inscription of a famous contemporary then very much alive:

Possessed of uncommon Powers for mimicking the Wailings of a Calf, the Brayings of an Ass, and other Musical Beasts, to the No little Entertainment of my Friends, and my own Satisfaction, how often have
I, J——s B——l, Esq.,
Descended from the Noble House of Bruce,
Now confined within these narrow Walls,
Been applauded for my Abilities, and considered
as concent'ring
In my own Person all the Characteristics of
The Animal Creation.

During the first two decades of the nineteenth century the various kinds of futuristic fiction found their characteristic patterns in imaginary wars, political prophecies, ideal states, science fiction stories, and ominous accounts of the Last Man. For every anxiety about the contemporary situation—the war against Napoleon, parliamentary reform, Catholic Emancipation—there was an appropriate narrative form in the tale of the coming invasion or in the political projection of future catastrophe. The possibility of a French sea-borne invasion, for example, appeared in the earliest accounts of future warfare; and the seeds of a minor publishing industry were planted in the melodramatic anticipations of defeat and victory in the anonymous projection of *The Invasion of England* (1803) and in William Burke's *The Armed Briton* (1806).

These dramas of the war-to-come, like the political projections, were popular exercises in pattern-building. The propagandists began with the known facts and with what they considered were real possibilities in the contemporary situation; and, in keeping with the calculated realism of their stories, they converted the whole of the immediate future into the image of a single and all too recognizable fear. This they enlarged to the limits of terror, so that anxieties about a French invasion or Catholic Emancipation would encourage the appropriate conclusion from the reader. Thus, the modest proposals for a more reasonable treatment of Catholic citizens had a sour response in *One Thousand Eight Hundred and Twenty-Nine* in which an anonymous Protestant patriot foresaw the return of the Inquisition and the restoration of the monastic lands to the religious orders. In a similar way the anonymous author of *The Red Book* warned the more liberally minded reader against the ideas of Sir Francis Burdett, the courageous parliamentary reformer. The argument by projection was that reform would lead to the disaster of a British Revolution: 'All the castles and country-houses of the nobility and gentry had been pillaged and burnt, and many of their proprietors were butchered with their servants and families.'

The exaggerated emotion, the feeling of impotence, the heightened sense of dread and the harrowing scenes so typical of this kind of futuristic fiction – these exhibit the qualities of the true nightmare. Indeed, this evident correspondence between human personality and group behaviour appears even more strikingly on those occasions when some major anxiety erupts into alarming accounts of the cataclysms and disasters that will descend upon the nation or the human race; and then there follows a period of collective neurosis when the future takes on the appearance of the dominant fear.

In their different ways these alarming visions illustrate the operation of two primary laws that control the tale of the future. First, and from the start, one of the most frequent themes has been the future not of individual nations but of all human beings and of the entire world. Second, there has always been a decided novelty about many of these projections, since writers find the substance of their hopes or fears in the most recent inventions and social changes. Today, propagandists begin with the fact of the Bomb, or ecology, computer developments, or African nationalism. And so it was in the beginning when Cousin de Grainville was writing the first chapters of *Le Dernier Homme* in 1798. He found his up-to-date ideas about the future of mankind in two publications of that year; and these proved immeasurably more effective than ever Mercier or Grainville could hope to be in making the idea of future change an accepted fact in contemporary thinking.

In the June of 1798 Dr Edward Jenner published the results of his research in a short pamphlet that gave the kiss of life to medical practice. This was the celebrated *Inquiry into the Causes and Effects of the Variolae Vaccinae*, which a committee of the House of Commons declared to be 'unquestionably the greatest discovery ever made for the preservation of the human species'. And as the world rejoiced at the discovery of vaccination, from the Tsarist court to the medical associations of New England, another writer broke in on the general happiness with a grim warning of the consequences of an ever-increasing population. According to Parson Malthus, as Marx called him, there would one day be too many mouths and not enough food. With his incisive phrases about geometrical and arithmetical ratios in the *Essay on the Principle of Population* Malthus revealed the cheerless future that waited for the citizens of the industrial societies at the very moment they were beginning to congratulate themselves on their conquest of nature. In a most ironic manner the predictions of Malthus and the discoveries of Jenner had one common effect. By concentrating attention on the way in which the sciences were changing the circumstances of human life, they made it abundantly clear that the future of mankind would be very different. The results were both general and particular. Jenner's vaccine strengthened the growing sense of expectation with the promise of a real change for the better in the condition of human life, and at the same time the Malthusian calculus made the world familiar with the practice of prediction.

Countless poems, congratulatory addresses, public awards and royal commendations poured in on Jenner; and all saluted him as the great benefactor who had changed the terms of the life-and-death struggle. The German medical journalist Kurt Sprengel wrote that Jenner had made 'the most important of all the discoveries of 1798 and the greatest of all eighteenth-century discoveries'. The poet Robert Southey saw Jenner as the liberator of mankind. 'That hideous malady which lost its power,' he wrote, 'When Jenner's art the dire contagion stay'd.' Even more, it brought the certainty of a happier future for the human race:

Fair promise be this triumph of an age,
When Man, with vain desires no longer blind,
And wise though late, his only war shall wage
Against the miseries which afflict mankind,
Striving with virtuous heart and strenuous mind
Till evil from the earth shall pass away.
Lo, this his glorious destiny assign'd!
For that blest consummation let us pray,
And trust in fervent faith, and labour as we may.[4]

Although Malthus did not attract the enthusiasm of the poets, his predictions started off a furious debate that raged for thirty years. His theories questioned the growing convictions of universal progress and prosperity; his conclusions rejected absolutely the hopes of Turgot, Joseph Priestley, William Godwin and the Marquis de Condorcet. As the first great controversy about the future population of the planet developed into a running battle of argument and counter-argument, the Malthusian theories became almost as well known as the campaigns of Bonaparte. There were six editions of the *Essay* in the lifetime of Malthus, many translations and many articles in which the Malthusians faced the attacks of Byron, Keats, Shelley, William Blake, Southey, Cobbett, Godwin, Hazlitt, Simon Gray and many others.

The main point at issue was the assertion in the *Essay* that 'man is necessarily confined in room. When acre has been added to acre till all the fertile land is occupied, the yearly increase of food must depend upon the melioration of the land already in possession. This is a fund, which, from the nature of all soils, instead of increasing, must be gradually diminishing.'[5] One of the most vigorous and perceptive answers came from William Hazlitt, who was very clear about the catalytic effect of the Malthusian proposition: 'He has not left opinion where he found it; he has advanced or given it a wrong bias, or thrown a stumbling block in its way.'[6] Malthus had shown that material progress had attendant problems, and he forced men to think of contemporary changes in terms of one world and of the one species *Homo sapiens*.

An immediate effect of the Malthusian doctrine was to raise

questions about the survival of the human race. The greatest calamity imaginable is the extinction of all life on our planet; and this possibility of an absolute end to human history is yet one more area of futuristic fiction for which writers have quarried their material out of the deeper levels of the psyche. From the earliest days of the tale of the future the experiences of the Last Man have displayed a continuing mythology of doom and desolation that has nowadays become a commonplace topic in science fiction stories. And here, once again, the French were the first in the field. The earliest account in fiction of the last days of mankind appeared in 1805 in *Le Dernier Homme* by Jean-Baptiste François-Xavier Cousin de Grainville; the same theme that later rang down the curtain on the nineteenth century in the Wellsian vision of a run-down solar system, when the Time Traveller flashed through the future 'in great strides of a thousand years or more, drawn on by the mystery of earth's fate, watching with a strange fascination the sun grow larger and duller in the westward sky, and the life of the old earth ebb away.'

This modern version of the archaic vision of an end to all things points to an eternal cycle of subliminal hopes and fears that reaffirm an apparent relationship between our doomsday stories and the ancient doctrines of the destroyed world in the Nordic *Ragnarok* and the Indian *Mahapralaya*. There is a striking similarity, for instance, between the final catastrophe in *Le Dernier Homme* and the North American myths of the cosmic disaster. According to the Cherokee Indians, 'when the world grows old and worn out, the people will die and the cords will break and let the earth sink down into the Ocean.'[7] According to Cousin de Grainville the Last Man will

> behold the plains and mountains stripped of verdure, dry and barren like a rock; the trees in decay and covered with a whitish bark; the sun, whose fire was grown dim, casting on every object a livid and gloomy light ... the earth had undergone the common destiny. After having for ages struggled against the efforts of time and men, who had exhausted it, she bore the melancholy features of decay.[8]

This *dies irae* for modern times came from a man of sixty who had begun his career in a seminary, the friend of the renowned Abbé Sièyes; and at the height of the Terror he had married in order to save himself from the guillotine. Grainville turned to teaching and authorship for his livelihood; and in the hard year of 1798, the year of Malthus, he began work on his epic history of the Last Man. Old, hungry and in desperate circumstances Grainville ended his life at two o'clock on the morning of February 1st, 1805, in the Somme canal at Amiens. Soon afterwards Sir Herbert Croft, the author of *The Abbey of Kilkhampton*, arrived in Amiens and asked to see Grainville, whom he called the author of the greatest prose epic of the day. On hearing that he was dead, the Englishman wept, protesting that had he known of Grainville's circumstances he would have saved him, asserting that the reputation of *Le Dernier Homme* would endure 'to the last man and to the end of the world'.

The world has long forgotten Cousin de Grainville, but the influences that caused him to write *Le Dernier Homme* are still as powerful as ever. The most important of them was the recurrent anxiety, so evident in the last 30 years, that comes from living in a period of unusually rapid and far-reaching changes – social, political, and technological. The other contributory factors were the traditional belief in the end of this world, literary ideas about the rise and fall of great nations, and the Malthusian teaching that there must be a limit to the growth of population. For example, in *Le Dernier Homme* the idea of the inevitable decline in all things joins with Christian belief and with Malthusian theory. The narrative opens with the appearance of the angel Ithuriel, 'the same who among the blooming bowers of Eden was the messenger of the Creator's high commands'. His mission is to prepare Adam, the Father of Men, for what 'will take place on the very day the destruction of the globe shall happen'. The end of human life is close:

> The inhabitants of the ancient world, after having exhausted their soil, inundated America like torrents, cut down forests coeval with creation, cultivated the mountains to their summits, and even exhausted that happy soil. They then descended to the shores of the ocean where fishing, that last resource of man, promised them an easy and abundant

> supply of sustenance. Hence, from Mexico to Paraguay, these shores of the Atlantic Ocean and South Seas are lined with cities inhabited by the last remains of the human race.[9]

God, it seems, had foreseen the theories of Malthus and the discoveries of Jenner. From the very beginning of mankind 'the duration of human life was wisely regulated by the omniscient mind of the Almighty, according to the size of the globe and the fecundity of its inhabitants.' And in the last days of humanity there was

> An additional motive for this limitation in the profound improvements making in medical science by which the lives of thousands of the infantine world have been snatched from the empire of death, and who, in thus becoming the heads of numerous progenies, are laying the foundation of an immense population which the earth in after-ages will be inadequate to sustain.[10]

The theme of the Last Man has certainly endured. It is a staple of much modern fiction about coming catastrophes, and it is a favourite subject for the cinema, which has produced many vivid projections of the last days of mankind. In *The Day the World Ended* (1955), *Dr Strangelove* (1963), and *The War Game* (1967) the perfect time-machine of the film presents the most recent images of the zero moment when history comes to an abrupt end. And all of them have their origin in Grainville's story and in the subsequent outpouring of stories, poems, and paintings that found the Last Man a worthwhile subject. In France a second edition of *Le Dernier Homme* came out in 1811 with an introduction from the original and learned writer, Charles Nodier; and there were several imitations: Auguste François Baron Creuze de Lesser, *Le Dernier Homme, poème imité de Grainville*, 1831; Alexandre Soumet, *La Divine Épopée*, 1841; Élise Gagne, *Omegar, ou le dernier homme*, 1858.

The more unusual developments of this theme were in poetry and in painting. Lamartine and Victor Hugo wrote about the desolate Paris of the future; and in England there was a vogue for poetry about the final catastrophe – Edward Wallace, Thomas Campbell, George Darley, Thomas Hood, and Byron all writing poems in which they described the end of life on earth. And it was undoubtedly a popular subject for the artists, who developed highly dramatic visions of chaos and ruin: John Martin composed a watercolour and an oil painting of 'The Last Man'; Joseph Gandy, the pupil of Sir John Soane who built the Bank of England, that symbol of eternity, made a watercolour of the bank in ruins; and in 1836 the American painter Thomas Cole produced four canvases on the theme of the decline and fall of empire. The source for these catastrophic visions was the English translation of *Le Dernier Homme* in 1806; and in their various ways they repeated the substance of the Grainville prophecy. Here is Thomas Hood:

> The sun had a sickly glare,
> The earth with age was wan.
> The skeletons of nations were
> Around that lonely man.

The same theme appeared in the closing chapters of Mary Shelley's *The Last Man* for which she borrowed many of her ideas from Grainville and from the poem 'Darkness' which Byron wrote in 1816:

> I had a dream, which was not all a dream.
> The bright sun was extinguish'd, and the stars
> Did wander darkling in the eternal space,
> Rayless, and pathless, and the icy earth
> Swung blind and blackening in the moonless air;
> Morn came and went – and came, and brought no day.

These visions of a last day for mankind appeared during the first two decades of the nineteenth century. At that unique time in history, when mankind was entering on a new way of life and the sciences were ending the ancient domination of nature, these prophecies of universal desolation were the imaginative renewal of a sacred bond between man and nature. At the time when the intellect was engaged in the sustained exploitation of the external world, the imagination had discovered within itself these catastrophic images that revealed grave anxieties for the future of humanity. These were in their way responses to the changing circumstances of urban life as valid as the contrary prophecies of human perfection that were then appearing in Europe. As belief or temperament inclined some writers to describe the worst of all possible futures, different ideas or a more sanguine frame of mind caused others to imagine the far happier condition of life in the coming centuries.

This chronic alternation between the extremes of despair and delight has affected the tale of the future from the start. The earliest indication of this clash between opposites in English fiction is in two projections that appeared within a year of each other. The first was *The Last Man* in which Mary Shelley found relief for the death of her husband by writing three volumes, beginning with the abdication of the last British monarch and ending with the plague that wipes out the inhabitants of Europe. A year later Jane Webb brought out a very different story, *The Mummy*, in which she found evident pleasure in her account of great technological advances.

During the first quarter of the nineteenth century this delight in the wonders of science induced some writers to indulge a liking for frolic and sport in the marvellous world of time-to-come. In 1800 the author of *Guirlanden um die Urnen der Zukunft*, A. K. Ruh, related a long romantic tale of a lost heiress and a devoted hero, the whole narrative located in the twenty-third century and enlivened with frequent escapades in dirigible balloons. Again, in 1803, Johann Daniel Falk produced a comic account of future air travel, *Elektropolis*, in which the passengers in an aerostat come down by parachute to the all-electric city, where the report has it that 'they eat by electricity, they drink by electricity, they sleep by electricity; and it is even said that the sun will soon be removed and an enormous electrophor put in its place.' The confident assumptions behind that ironic prediction show how the *Zeitgeist* had found a new means of revelation in the *Zukunftsroman*; and this fact of the new futuristic fiction becomes clear beyond all doubt in the preface to a romantic story of 1810, *Ini: ein Roman aus dem ein und zwanzigsten Jahrhundert*: 'As assuredly as the present is an improvement on the past, so with equal certainty a better future is coming; and we may at least be confident that we can expect our ever-developing civilization will be the salvation of all mortal creatures. What we cannot yet see, we dream of.'[11] The message of hope came from a Prussian army officer, Julius von Voss, whose tale of the twenty-first century has earned him a special place in the history of futuristic fiction, since *Ini* was the first story to combine extensive political changes with all the magical effects of an early science fiction technology.

The scene is set in a world transformed – the Emperor of Europe rules over one united dominion from Sicily to Moscow; the Empire of New Persia controls all Asia less China; and after losing India to the Persians, the Albionen have settled in Australia. The action is for the most part a tale of love and adventure – the hero does not know he is the son of the Emperor of Europe, and in a spirit of true devotion he sets off for military service only to find at the end of the narrative that he has been betrothed to the girl he had loved from the start, Ini, the daughter of the Empress of Africa.

The vast scale of the action – journeys to America and the North Pole – was the consequence of an equally extensive series of technological marvels that make this German author a Jules Verne before his time. In the advanced world of the twenty-first century the citizens of Europe can travel in balloons towed by eagles, or on floating islands towed by whales, or along special horse-ways in carriages with house-high wheels. A rapid postal service operates by means of cannon that shoot the mail in special containers from town to town. Berlin has become a sea city; they have discovered the North Pole and the secret of perpetual motion; there are synthetic diamonds, vast underground cities with artificial lighting, diving machines with special equipment for submarine work. The nations of the future fight out their wars with a variety of new weapons: air machines drop inflammable material on warships; naval artillery fires incendiary shells at the enemy; and submarine troops attack naval craft with waterproof explosives.

The style and the content of these early German stories and the evidence of similar productions in English and in French make it abundantly clear that writers in Europe – and later in the United States – had found the new tale of the future an ideal means of commenting on or looking into the possibilities suggested by the many changes that were taking place during the first decades of the nineteenth century. Everywhere there was a great uniformity in narrative methods, and these always depended on those dynamic factors that have decided the workings of the tale of the future from *L'An 2440* to the most recent doomsday story: a marked feeling of total freedom – a sense of the boundless – that allows the imagination to stray as it pleases through all times and all lands, through the whole world and even to the remotest parts of the solar system; an evident pleasure of the mind that encourages the characteristic readiness to consider every idea and every kind of possibility within the bounds of contemporary knowledge; an undoubted certainty about the process of change in human society that sees the future as one more stage, for good or ill, in the continuing evolution of mankind; and, in many ways the most powerful factor of all, a dominant mood that subordinates everything to the excitement of an adventure story, or the calculations of an admonitory forecast, or the revelation of life in an ideal commonwealth, or the consequences of some social or natural catastrophe.

All this is to say that the tale of the future was the immediate product of the new technologies and of the new attitude to history that emerged during the second half of the eighteenth century. The one acted on the other to encourage a habit of mind that sought to derive the shape of the future from the realities of contemporary experience; for the tale of the future had the double advantage that technological innovation had made prediction an everyday practice, and that in the prevailing historicism of the age it was as natural for the Duc de Lévis to look ahead to the Europe of 1910 in his *Voyages de Kang-hi* as it was for Sir Walter Scott to find the essential character of the nation in the record of the past.

Wordsworth had these facts in mind when he wrote the preface to the second edition of the *Lyrical Ballads* in 1800. He pointed to the effects of 'a multitude of causes, unknown to former times ... The most effective of these causes are the great national events which are daily taking place, and the increasing accumulation of men in cities.' And after he had developed his theory of the similarities and the differences between the poet and the scientist, Wordsworth began to prophesy in the manner of the times:

> If the labours of men of science should ever create any material revolution, direct or indirect, in our condition, and in the impressions which we habitually receive, the Poet will sleep no more than at present; he will be ready to follow the steps of the Man of science, not only in those general indirect effects, but he will be at his side, carrying sensation into the midst of the objects of science itself. The remotest discoveries of the Chemist, the Botanist, or Mineralogist, will be as proper objects of the Poet's art as any upon which it can be employed.[12]

Wordsworth belonged to the first generation of the new age. He was born in the year before the publication of *L'An 2440* and he died in the year before the Great Exhibition. Throughout his long life his poetry had recorded the feelings of an Englishman during a period of turmoil and upheaval – from the jubilant early days of the French Revolution, 'when Reason seemed the most to assert her rights', to his prophetic celebration of technological achievement in the sonnet 'Steamboats, Viaducts, and Railways':

> Nor shall your presence, howsoe'er it mar
> The loveliness of Nature, prove a bar
> To the Mind's gaining that prophetic sense
> Of future change, that point of vision, whence
> May be discovered what in soul ye are.
> In spite of all that beauty may disown
> In your harsh features, Nature doth embrace
> Her lawful offspring in Man's art; and Time,
> Pleased with your triumphs o'er his brother Space,
> Accepts from your bold hands the proffered crown
> Of hope, and smiles on you with cheer sublime.

Wordsworth wrote in the idiom of the new literature of the future – time, space, art, triumph, hope, vision – and his most revealing phrase about 'that prophetic sense of future change' was an exact definition of the universal expectancies that had brought the tale of the future into service as the appropriate response in European fiction to the transformation of society.

The experiences recorded in the German *Zukunftsroman* find confirmation in the French and English projections that appeared during the lifetime of Wordsworth. In 1802 the very original and lively writer Restif de la Bretonne published *Les Posthumes*, in which he found pleasure in playing with time and took the opportunity to let his narrative follow imaginative speculation wherever it pleased. The major invention in *Les Posthumes* is the Duc de Multipliandre who plays the part of a benevolent Faustus. More fortunate than Frankenstein, Multipliandre used his knowledge of science to turn himself into the first Superman in fiction. He can exchange his intellect with anyone at any time, and by this power he is able to take control of Frederick II of Prussia, the Emperor Joseph of Austria, and the Pope – with astonishing effects on the course of history. Long before H. G. Wells thought of the idea, this Superman of 1802 had found the secret that made him the first Invisible Man in the history of science fiction – the result of a chemical formula in which 'the negative elements prevent the body from receiving any ray of light. I was able to make up the mixture thanks to Lavoisier', the text notes with careful authentication, 'who gave me the materials of which I only knew the names.'

Fortunately for the world this self-made Prometheus is virtue itself. He devotes many centuries of his immortality to reforming the nations and establishing a world government; he travels through the solar system as far out as Jupiter and Saturn; and from the centuries ahead the time-traveller reports that a new race of human beings will emerge after a comet causes a world flood. And finally, again long before H. G. Wells, this vision of the remote future passes on the ominous prediction that in the year 99,756 the Earth is moving closer to the Sun.

After that explosion of the imagination there can be no doubt that the tale of the future is the dreamtime of industrial society. It reaches down to the mythic roots within human experience to find sources of supreme power, means of transcending all limitations, opportunities for achieving absolute perfection. Like the alphabet, the tale of the future is a necessary social invention. It provides an imaginative and conjectural code that makes it possible for any writer to describe any apprehension or any anticipation between the extremes of nightmare and ecstasy; it offers a plastic and visual image that can take on the shape of any social theory, mirror any possibility, create any kind of feasible world; and as the spirit of prophecy moves the writer, so the imagined future can fall into any pattern, from the extraordinary Prospero-visions in *Les Posthumes* to commonplace descriptions of the next stage in technological achievement.

For example, the Duc de Lévis wrote into *Les Voyages de Kang-hi* of 1810 a number of forecasts that came hot from the assumptions of his time. His Chinese time-traveller reports that 'the extermination of smallpox and the progress of agriculture have increased the population by one third.' He notes the advances that have taken place by the beginning of the twentieth century – a canal from Lyons to the sea, railways between Paris and Orléans, a canal from Suez to Alexandria, air-conditioning in the homes of the wealthy, and the invention of the *melodico-humana*, which is the gramophone of the future. The author is so certain of the constant rate of development in society that he includes the first imaginary newspaper of the future, *Le Journal du déjeuner*, 15 Septembre 1910, an ingenious device for demonstrating the similarities and differences between the Europe of 1810 and the changed world of the twentieth century.

The historians of the future discovered the configuration of the coming centuries in the journals and newspapers of their time. In 1828, for instance, the author of *A Hundred Years Hence* began a chapter with the seemingly casual observation:

> I had promised a friend of mine (who has a magnificent chateau, par parenthèse, not far from Grantham upon the high north rail-road) to go and pass two or three days with him, and as I left London before noon, I was proceeding very leisurely along, at the rate of twenty miles an hour only, in my pocock with a pair of kites, when all of a sudden I perceived a gentleman's light steam-barouchette ...[13]

The style is the essence of futuristic fiction, for the projected world of the twentieth century depends for its acceptability on facts and assumptions that are known to derive from contemporary knowledge. Thus, the description of the 'pocock with a pair of kites' was an up-to-date borrowing from the experiments of George Pocock who had described his scheme for a new method of kite-powered transport in a publication of 1827, *The Aeropleustic Art of Navigation in the Air by the Use of Kites or Buoyant*

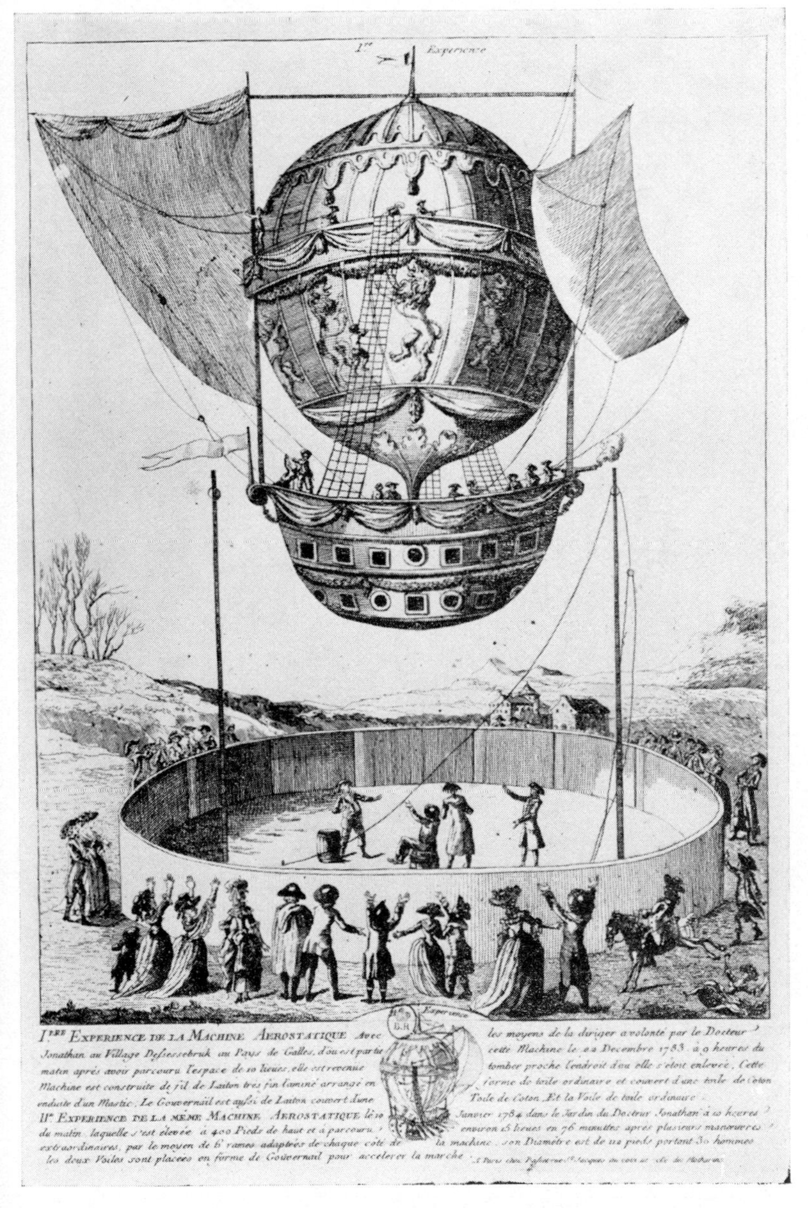

1 From the start the new images of the future discovered their shape in the emergent technologies of their time. The potential suggested the fantastical.

2–3 Whatever the possibility – the danger of a French invasion, for example, or the use of balloons for air operations – the artists had an image for it.

4 The ever-growing knowledge of the past in the first half of the nineteenth century was an element in the new idea of progress.

5 And the idea of progress found validation in the discoveries of the palaeontologists.

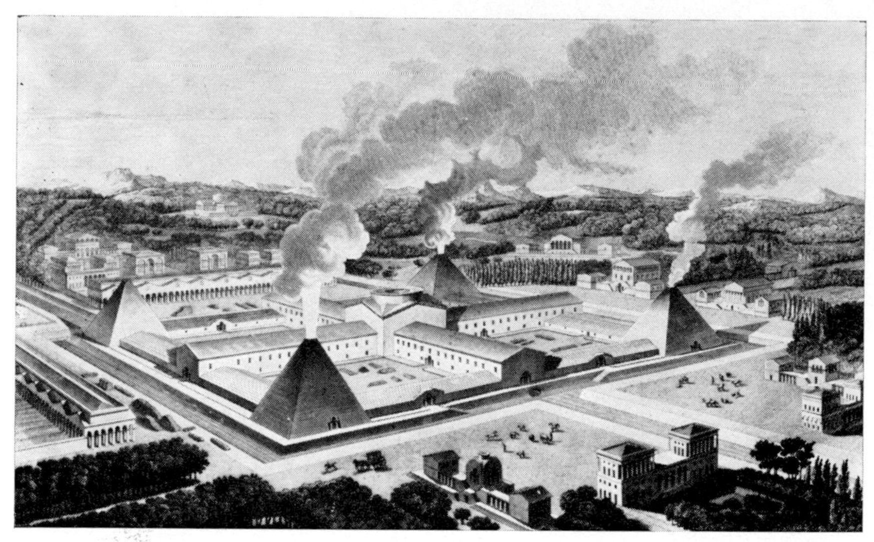

6 The vision of the future could be a scheme for an advanced forging establishment in 1804.

7 And it could be a genial frolic with the fantastic in a popular print.

8–9 The whimsicalities of these popular prints of 1828 display that pleasurable playing with the possible—which is the origin of many futuristic fantasies.

10 The celebrated 'Lunar Hoax' of 1835 was an opportunity for European artists to try their hands at depicting the Moon Man supposed to have been seen by Sir John Herschel.

11 In a similar way the reports of Henson's experiment with a model flying machine in 1843 produced numerous prints of a heavier-than-air machine in flight.

12 Writers of fiction looked forward to the day when men would be able to go to the Moon by means of anti-gravitational devices.

13 Jules Verne began from the engineering calculations about the power required to lift a flying machine into the air.

14 The powers of science to the rescue of human helplessness – a telling image from Jules Verne's *Robur le Conquérant*, 1886.

Sails. Again, the reference to the steam-barouchette and a later reference to 'Mr Gurney's steam carriages' came from another borrowing, this time from the work of the Cornish engineer Sir Goldsworth Gurney, who began experiments with a steam-driven vehicle in 1822. These went so well that Gurney had the glory of making the first long journey in a self-propelled vehicle from London to Bath and back in the July of 1829.

The forecasts in *A Hundred Years Hence*, like the later science fiction stories of Jules Verne, were delightful ways of projecting the desirable. In those first decades of the last century, when the idea of progress was becoming established doctrine, it was part of the common experience to think and talk about the future in the language of science fiction. In 1819 the renowned explorer of the past, Baron Georges de Cuvier, told the Institut royal de France of likely innovations in steam navigation: 'What we can see in the future, and what will be perhaps even more important, is the change that steam navigation will bring about in maritime warfare and in the power of nations. It is extremely probable that in this we will see another of those advances that can be included in the list of those developments that have changed the face of the world.'[14] On December 20th, 1825, the principal citizens of Glasgow met in the hall of Anderson's College, the first technological foundation in the history of the United Kingdom, and heard the president speak on the proposals for a monument to James Watt. To the applause of the audience, he ended on a note of pure science fiction, telling them that 'the time is not far distant when chariots winged with fire shall be seen flying over metallic pavements through all the populous districts of the empire, transporting travellers and merchandise with amazing smoothness and velocity.'[15]

By the 1820s conjectures about the future had become as familiar as the forecasts of the futurologist have been in the 1970s. Editors used occasional pieces of the kind that appeared in *The Literary Melange* for August 1822, 'Specimen of a Prospective Newspaper', which gave the reader an amusing account of life in the forty-eighth century. And it was no doubt a sign of the times in 1824, when the young Macaulay became a Fellow of Trinity College, Cambridge, and wrote one of his first articles for *Knight's Quarterly Magazine*. This was 'A Prophetic

Account of a Grand National Poem to be entitled "The Wellingtoniad", and to be published A.D. 2824', the whole presented in the future-comic style then in favour with the magazines.

These trifling comedies of a later age had their origin in a complacent attitude to the future, and this in turn sprang from the general awareness of change which had domesticated the idea of progress within the new literature of prophecy. Throughout the emergent industrial nations it was a central belief that science had made possible a new compact with nature; and in consequence the extrapolation of future developments took on decidedly utopian qualities in those forecasts that looked forward to the transformation of the world. In 1827 the Reverend Dionysius Lardner brought out one of the earliest histories of the steam-engine, and in his preface he stated without any fear of contradiction that Watt's invention had

> ... increased the sum of happiness, not only by calling new pleasures into existence, but by so cheapening former enjoyments as to render them attainable by those who before never could have hoped to share them. Nor are its effects confined to England alone; they extend over the whole civilized world; and the savage tribes of America, Asia, and Africa must ere long feel the benefits, remote or immediate, of this all-powerful agent.[16]

This was common doctrine. In the following year the members of the Institution of Civil Engineers wrote the convictions of the age into their petition for a charter of incorporation: 'Civil Engineering is the art of directing the great sources of power in Nature for the use and convenience of man; being that practical application of the most important principles of natural philosophy which has, in a considerable degree, realized the anticipations of Bacon.'[17]

Those words were far more than a simple statement of pride in the extraordinary achievements of the engineers. The studied reference to 'the anticipations of Bacon' proclaims a new-made collaboration between the ideal and the technological. Again, the utopian phrase about 'the great sources of power in Nature' sums up what many authors of ideal states of the future had been saying with increasing clarity ever since 1771. All the evidence shows that from the beginning, in Mercier's *L'An 2440* and the 'Essay on Population' by Malthus, the spirit of prophecy and the business of prediction had worked in parallel to detect and describe the course of future developments. In 1808, for example, the original and somewhat eccentric French ideologue Charles Fourier wrote of the industrial armies that would one day change the face of the world:

> They will execute works the mere thought of which would freeze our mercenary souls with horror. For instance, the combined order will undertake the conquest of the great desert of Sahara; they will attack it at various points by ten and twenty million hands if necessary; and by dint of transporting earth, cultivating the soil and planting trees every here and there, they will succeed in rendering the land moist, the sand firm ... They will construct canals navigable by vessels, where we cannot even make ditches for irrigation, and great ships will sail not only across isthmuses like those of Suez and Panama, but even in the interior of continents, as from the Caspian Sea to the Sea of Azov, of Aral, the Persian Gulf; they will navigate from Quebec to the five great lakes.[18]

As the rate of development accelerated and the forecasts multiplied, those who read the journals took it for granted, like the young Tennyson, that men would one day conquer the air and that engineers would link the Indian Ocean with the Mediterranean and drive a canal across the Isthmus of Panama. In 1827, for instance, Jane Webb published an exuberant Gothic romance, *The Mummy*, in which she described the technological wonders of the twenty-second century: steam mowing apparatus, electrical rain-making equipment, steam digging-machines, balloons that rise to a height of seventeen miles and mobile houses that move from place to place on railway lines. Her prophecy of the new Egypt makes her the first sibyl in the history of modern technology. The balloonists see a new world beneath them:

> Different, however, oh! how different from the Egypt of the

> nineteenth century, was the fertile country which now lay beneath their feet. Improvement had turned gigantic steps towards its once desert plains; commerce had waved her magic wand; and towns and cities, manufactories and canals, spread in all directions. No more did the Nile overflow its banks: a thousand channels were cut to receive its waters. No longer did the moving sands of the Desert rise in mighty waves, threatening to overwhelm the way-worn traveller: macadamized turnpike roads supplied their place, over which post-chaises, with anti-attritioned wheels, bowled at the rate of fifteen miles an hour. Steamboats glided down the canals, and furnaces raised their smoky heads amidst groves of palm-trees; whilst iron railways intersected orange groves and plantations of dates and pomegranates might be seen bordering excavations intended for coal pits. Colonies of English and Americans peopled the country, and produced a population that swarmed like bees over the land, and surpassed in numbers even the wondrous throngs of the ancient Mizraim race; whilst industry and science changed desolation into plenty, and had converted barren plains into fertile kingdoms.[19]

In the same year of 1827, when *The Mummy* was on the shelves of the circulating libraries, Goethe had read of the proposals made by the German geographer Alexander von Humboldt for a canal between the Atlantic and the Pacific. At dinner on February 21st he told the ever-faithful Eckermann of his hopes for the future, and the Boswell of German literature reports Goethe as saying:

> All this is reserved for the future and for an enterprising spirit. So much, however, is certain that, if they succeed in cutting such a canal that ships of any burden and size can be navigated through it from the Mexican Gulf to the Pacific Ocean, innumerable benefits would result to the human race. But I should wonder if the United States were to let an opportunity of getting such a work into their own hands escape. It may be foreseen that this young state, with its decided predilection to the West, will in thirty or forty years have occupied and peopled the large tract of land

> beyond the Rocky Mountains. It may furthermore be foreseen that along the whole coast of the Pacific Ocean, where nature has already formed the most capacious and secure harbours, important commercial towns will gradually arise for the furtherance of a great intercourse between China and the East Indies and the United States ... Would that I might live to see it!—but I shall not. I should like to see another thing—a junction of the Danube and the Rhine. But this undertaking is so gigantic that I have doubts of its completion, particularly when I consider our German resources. And thirdly, and lastly, I would wish to see England in possession of a canal through the Isthmus of Suez. Would I could live to see these three great works! It would be worth the trouble to last some fifty years more for the very purpose.[20]

In this way did the idea of the future enter into the general experience of the educated, so that by the 1830s the tale of the future had become an established literary form in Europe. As the simpler expectations of the eighteenth century vanished in the torrent of inventions and discoveries, the increasing confidence and subtlety of the new fiction was the outward sign and immediate consequence of a far more comprehensive and decided attitude to time. One of the major factors in this movement of ideas was the rapid advance in the knowledge of the prehistoric past during the first half of the nineteenth century; and this new information about the immense sequence of development from the Palaeozoic to the Palaeolithic joined with the new evolutionary theories to give the idea of progress the authority of a universal law. Another equally important factor was the work of influential historians—Guizot, Macaulay, Taine, Lecky, Buckle—who explained the operation of the necessary, causal laws in the Newtonian universe of human society; and at their most sanguine they proclaimed the comforting theorem that the onward march of civilization would continue into the infinite future. The proofs seemed all about them—in the rapid growth of the railways, the new steamships, and the dazzling colours that followed on the discovery of the first synthetic dyes.

As fact linked with fact to indicate the evolutionary design in nature, and as invention followed invention in the incessant exploitation of the sciences, the language received the surge of discovery in a flood of new words. In the 1830s the geologists began to mark off the remote past in millions of years—*Cambrian*, *Silurian*, *Jurassic*, *Eocene*, *Miocene*, *Pliocene*; and the science of palaeontology, a new word in 1838, drew on a dead language to name extinct forms of life—*echinoderm*, *brachiopod*, *calamite*, *iguanodon*, *ichthyosaurus*, *dinosaur*. From the laboratories and the factories came word of new machines, new methods, new materials—*creosote*, *aniline*, *gutta-percha*, *stethoscope*, *chloroform*, *electrolysis*, *electric light*, *commutator*, *transatlantic*. And in this way the dictionaries recorded the history of progress from the first entry of *aeronaut* in 1784 to the most apt appearance of *carphology* in the year of the Great Exhibition of 1851.

These advances and discoveries promoted the new attitude to time that encouraged all to look upon the past as a panorama of evolutionary change and on the future as a scenario that could reveal every imaginable kind of development. In 1829 the poet and biographer Robert Southey summed up the experiences of his lifetime in the first major study of the idea of progress in the English language, *Sir Thomas More; or, Colloquies on the Progress and Prospects of Society*; and in his review of that book Thomas Babington Macaulay clearly felt it was the most natural thing in the world for him to look into the future as far as he could see and

> ... to prophesy that in the year 1930 a population of fifty millions, better fed, clad, and lodged than the English of our time, will cover these islands, that Sussex and Huntingdonshire will be wealthier than the wealthiest parts of the West Riding of Yorkshire now are, that cultivation, rich as that of a flower-garden, will be carried up to the very tops of Ben Nevis and Helvellyn, that machines constructed on principles yet undiscovered will be in every house.[21]

That will-be was the point of vision from which the prophets of the last century were able to discern their new Eden of mechanical invention and improved living conditions. That readiness to prophesy had become a habit of the age. As the theories of progress and the new philosophies of history systematized the universal experience of change, the many proposals for future developments – from Stephenson, Brunel, Ferdinand de Lesseps – provided an education in the practice of extrapolation and forecasting. The result was a style of writing and thinking about progress that was common to the tale of the future, to extrapolations of the expected advances in technology and society, and to the general discussion of the achievements of the age.

When the French astronomer Dominique François Arago delivered the customary *Éloge* to the Académie des Sciences in commemoration of the death of James Watt, he chose to present their most illustrious Foreign Associate as the first hero of progress toiling for ever in the eternal future of a world transformed:

> Let us reckon upon the future! A time will come, when the science of destruction shall bend before the arts of peace ... Then Watt will appear before the grand jury of the inhabitants of the two worlds. Every one will behold him, with the help of his steam-engine, penetrating in a few weeks into the bowels of the earth to depths which, before his time, could not have been reached without an age of the most toilsome labour; excavating vast mines, clearing them in a few minutes of the immense volumes of water which daily inundated them ... The population, well supplied with food, with clothing and with fuel, will rapidly increase; it will by degrees cover with elegant mansions every part of the earth, even those which might justly have been termed the Steppes of Europe, and which the barrenness of ages seemed to condemn to be for ever the exclusive domain of wild beasts. In a few years hamlets will become great towns; in a few years boroughs, such as Birmingham, where there could hardly be counted thirty streets, will take their place among the largest, the handsomest and the richest cities of a mighty kingdom.[22]

This enthusiastic language was the most favoured means of describing the facts of progress and the future. As the title of a American forecast of 1836 promised, the world of the future was to be *The Paradise within the Reach of All Men without Labour, by Powers of Nature and Machinery*; and in rhapsodic phrases the author, J. A. Etzler, foretold how science would change the world. The great powers of science, he argued,

> ... are more than sufficient to produce a total revolution of the human race, as soon as understood; for they can effect in one year more than hitherto could be done in 10,000 years, and things unheard of. The world will take a quite different appearance than it has had hitherto to man; productive of a thousand times more means for human happiness, than the human race may be wanting; a paradise beyond the common conceptions.

The prophet described how it would be possible to exploit nature in the service of mankind, and he closed his revelations with a vision of the technological Jerusalem that would emerge in the heaven of the future:

> Palaces, superior in magnificence, grandeur and commodities to anything known; of crystal-like appearance inside and outside, and indestructible for thousands of years; constructed as if of one entire piece, for the common habitations of men everywhere; floating islands of light wooden stuff impervious to water, made of any kind of wood, covered with fertile soil, bearing trees and all kinds of desirable vegetables, with palaces and gardens, and thousands of families for their inhabitants, exempt from all dangers and incommodities; which may move by powerful engines at the rate of 1000 miles per day through the ocean. Men may travel over land and sea from pole to pole in a fortnight; from America to Europe in three or four days with a certainty. All things desirable for human life, when once known, may be rapidly multiplied without labour or expense to superabundance for everyone ... The same powers may also be used as weapons for conquering and subjecting nations, because they afford means to which neither gunpowder nor armies of any number of men can resist. Europe will be approached to America within three or four days' journey by means of impregnable fortresses. The question will hence be whether America or Europe is to be the ruling power.[23]

This artless rhapsodizing gave notice that a new and much more self-assured generation had taken on the mantic role of Condorcet and Sebastien Mercier. Historians and social theorists as different as Macaulay and Marx, Comte and Buckle, shared Tennyson's conviction that '... thro' the ages one increasing purpose runs / And the thoughts of men are widen'd by the process of the suns.' The idea of progress was one of the potent beliefs of the age. It remains to be seen how it found a most popular means of communication in the prophecies, projections and predictions of futuristic literature.

3

The Prospect of Probabilities

By the 1830s the idea of the future had become a familiar element in contemporary thinking throughout the industrial nations. The prophecies and projections of futuristic fiction had their basis in an unquestioning readiness—indeed, a general eagerness—to believe that to the wonder-working sciences all things were possible.

Twenty years before the first stories of Jules Verne and fifty years before Wells wrote *The Time Machine*, the imagination had found material for romance in a prospect of probabilities—rational life on the Moon, advanced mechanical contrivances, world communications—and these have developed over the years into the space stories, the robots and androids and technological wonders of modern science fiction. From the origination of the new fiction the subtle engines of the will-to-believe have operated with uniform effect on both sides of the Atlantic. The contemporary mythology of flying saucers, for example, has an early-nineteenth-century precedent in the report of flying creatures on the Moon that appeared in the New York *Sun* during the August of 1835. This notorious episode of the Great Lunar Hoax displays that artful combination of fact and fantasy that supplies both sanction and substance to the ingenuities of science fiction.

First, there were the facts of science. On January 15th, 1834, the eminent astronomer Sir John Herschel arrived at Cape Town, where he intended to observe the constellations of the southern latitudes; and on March 5th, preparations completed, he turned his 18-inch telescope on the Southern Cross. Second, there were the facts of commercial enterprise as they appeared on August 21st, 1835, to Benjamin H. Day, founder of the New York *Sun*, when Richard Adams Locke outlined his scheme for a completely fanciful account of Herschel's discoveries. At that time Benjamin Day was eager to exploit any means of increasing the sales of the *Sun*; and Locke, English immigrant and able journalist, was looking for ways of adding to his weekly salary of twelve dollars. On August 25th the two men began business with three columns across the first page of the *Sun* which promised the readers in double pica headlines the first account of the 'Great Astronomical Discoveries lately made by Sir John Herschel, LL.D., F.R.S., &c. at the Cape of Good Hope.' In the solid fashion of honest journalism Locke opened by quoting (with full acknowledgments) the evidence of a special supplement on Herschel that had appeared in the *Edinburgh Journal of Science*. The readers, who did not know that the journal had ceased publication in 1828, had good reason to accept what science told them:

> In this unusual addition to our journal, we have the happiness of making known to the British public, and thence to the whole civilized world, recent discoveries in astronomy which will build an imperishable monument to the age in which we live, and confer upon the present generation of the human race a proud distinction through all future time. It has been said that 'the stars of heaven are the hereditary regalia of man' as the intellectual sovereign of the animal creation. He may now fold the zodiac around him with a loftier consciousness of his mental supremacy.[1]

Day by day Locke convinced his readers by brow-beating them with the scientific facts he claimed to have found in the *Journal of Science*. There was a detailed but spurious explanation of the great reflecting telescope, learned talk about the properties of light, theories about the possibility of life on the Moon, references to the Royal Society and the British Board of Longitude – all part of Locke's careful preparation for the evening of January 10th, when the astronomer turned his telescope to the lunar disc. Then, to reinforce the sense of realism, Herschel's assistant supposedly took up the narrative and gave an eye-witness account of what they saw. 'Small collections of trees, of every imaginable kind,' wrote Dr Andrew Grant,

> were scattered about the whole luxuriant area; and here our magnifiers blest our panting hopes with specimens of conscious existence. In the shade of the woods, on the south-eastern side, we beheld continuous herds of brown quadrupeds, having all the external characteristics of the bison, but more diminutive ... The next animal perceived would be classed on earth as a monster. It was of bluish lead-colour, about the size of a goat, with a head and beard like him, and a *single horn*, slightly inclined forward from the perpendicular.[2]

The story moved on, the scientist reporting the discoveries with copious scientific details; and then came the astounding news that there were rational beings on the Moon:

> We counted three parties of these creatures, of twelve, nine, and fifteen each, walking erect towards a small wood near the base of the eastern precipices. Certainly they *were* like human beings, for their wings had now disappeared, and their attitude in walking was both erect and dignified ... They averaged four feet in height, were covered, except on the face, with short and glossy copper-coloured hair, and had wings composed of a thin membrane, without hair, lying snugly upon their backs, from the top of the shoulders to the calves of the legs.[3]

Finally, before Locke came to the stupendous discovery of buildings and temples on the Moon, he prepared the reader in his astute way by revealing that Herschel had removed several passages from Dr Grant's narrative:

> From these, however, and other prohibited passages, which will be published by Dr Herschel, with the certificates of the civil and military authorities of the colony, and of several Episcopal, Wesleyan, and other ministers, who, in the months of March last, were permitted, under stipulation of temporary secrecy, to visit the observatory and become eye-witnesses

> of the wonders which they were requested to attest, we are confident his forthcoming volumes will be at once the most sublime in science, and the most intense in general interest, that ever issued from the press.[4]

The Locke story certainly had a part in the development of the press in the United States. The double-cylinder Napier machine in the *Sun* basement ran at full speed ten hours a day to meet the demand for additional copies; and on August 28th, when the *Sun* sold 19,360 copies, Benjamin H. Day boasted with honest Yankee pride that his paper had surpassed the 17,000 daily average of the London *Times* and had thereby created a world record for the largest circulation of any daily newspaper anywhere.

Edgar Allan Poe, who had a professional regard for the good hoax, claimed that the Moon story had been a major factor in promoting the mass-circulation press:

> From the epoch of the hoax 'The Sun' shone with unmitigated splendour. The start thus given the paper insured it a triumph; it has now a daily circulation of not far from fifty thousand copies, and is therefore, probably, the most really influential journal of its kind in the world. Its success firmly established 'the penny system' throughout the country, and (*through* 'The Sun') consequently, we are indebted to the genius of Mr Locke for one of the most important steps ever yet taken in the pathway of human progress.[5]

As the news from the Moon spread across the United States, the majority of readers showed their eagerness to believe by spending their money on the pamphlet editions that poured out by tens of thousands from the hard-worked Napier printing machine. Two professors from Yale (Denison Olmsted and Elias Loomis) sped down the Hudson in search of the pure knowledge of the *Journal of Science*, which they thought was kept in the *Sun* editorial office; and Harriet Martineau, then travelling west through Massachusetts, found that the ladies of Springfield were collecting for a fund to send missionaries to the Moon.

Later on, when copies of the *Sun* arrived in Europe, an even larger audience took to the Herschel story with equal enthusiasm. Throughout the year of 1836 the translations multiplied. There were editions in French, German, Danish, Italian, and from Llanrwst in the remote hills of North Wales came an illustrated translation that confirmed the universal interest in *Hanes y Lleuad; yn Goson Allan y Rhyfeddodau a Ddarganfyddwyd gan Syr John Herschel.* In France the Director of the Paris Observatory, Dominique François Arago, brought the hoax to the notice of the Académie des Sciences in order to protect the reputation of his friend, Sir John Herschel. In Italy there were several illustrated versions of the 'telomicroscopic news' from the Moon; and the best of these came from Naples, where two gifted engravers – Gaetano Dura and Enrico Fergola – must have done very well for themselves with the sets of engaging lithographs they produced in the April of 1836. The prize for originality, however, must go to the French; for the friend and disciple of Fourier, Victor Considérant, made cunning use of the Lunar Hoax in order to demonstrate the happiness of the phalansterian social system. Considérant produced an enlarged version of the Locke story. He began with a humble dedication from Herschel to the King of England ('I am happy, Sire, to be able to place at the feet of Your Majesty the results of our long labours') and to the original text Considérant added long passages in which he described the splendid Fourierist buildings of Selenopolis and made well-informed conjectures about the social habits of the Selenites.

The astonishing diffusion of the Lunar Hoax was, in the first place, the result of improved communications and steam printing-machines, whereas the international response to the story was a consequence of the new mythology that had conditioned the peoples of two continents to accept whatever came to them with the authority of the sciences. As the propaganda application of the Considérant version demonstrates, readers in Europe had found their expectations realized in the Locke story. Because it seemed to them that the growth of scientific knowledge had defined the direction and described the shape of future developments, they welcomed the pseudo-astronomical account of flying creatures and lunar cities as a signal that they had moved on yet another stage in the progress of mankind.

The possibility of life on other planets had been a matter for discussion amongst the educated ever since Fontenelle published the *Entretiens sur la pluralité des mondes* in 1686, and during the early decades of the nineteenth century several popular books on astronomy had encouraged speculation on the subject. In the judgment of William Griggs, who published an account of the Lunar Hoax in 1852, these prepared the readers of 1835 for the *Great Astronomical Discoveries*:

> So thoroughly was the popular mind, even among the best educated and most reading classes, imbued with these fanciful anticipations of vast lunar discoveries, and of speedy telescopic improvements by which they were assuredly to be achieved, that, at the time Mr Locke's 'Moon Story' was written, scarcely anything could have been devised and announced upon the subject too extravagant for general credulity to receive.

'Credulity' is too harsh a word to describe the whole-hearted belief in the powers of science so characteristic of the nineteenth century. As Sir John Herschel told a London audience in 1830, speculation is the attribute of a progressive age; and for that reason, whenever a man considers the power of science, 'he is no longer content to limit his enterprises to the beaten track of former usage, but is constantly led onwards to contemplate objects which in a previous stage of his progress, he would have regarded as unattainable and visionary'.[6] That was the belief of the enthusiastic and energetic Saint-Simonian Barthélemy Enfantin, who went to Egypt in 1833 with proposals for a Suez Canal, a Cairo-to-Suez railway and a Nile barrage. And the same readiness to consider any possibility can be seen in the preface that the German astronomer and mathematician Joseph Nürnberger wrote for *Die Seleniten*, a lunar travel story of 1833. Of all the branches of astronomy, he thought, there could be 'none more exciting for the imagination than the so-called topography of the heavens. What the telescope discovers about the nature of the other celestial bodies becomes an apt subject for conjecture.' The wonders of the star-lit night, he went on, encouraged a sense of expectation and a lunar fantasy 'brings us on the wings of the imagination to the meadows of the Moon, showing us in vivid pictures not only what is there but also what can and might be there.' That might-be is the modal core of all futuristic fantasies about other worlds in space; and the author showed that he was well aware of the necessary link between contemporary knowledge and projected possibilities, since he devoted the whole of the first chapter to explaining how a phlegmatic English nobleman, Lord Eaglespeed, took off for the Moon on August 8th, 1930, in the very latest of the new-model Montgolfier balloons. And here the coincidence of this story and the Lunar Hoax shows the working of a primary principle in the tale of the future—that the facts must always precede the fantasy. Before the imaginative descriptions in *Die Seleniten* there were the conjectures of the eminent astronomers Wilhelm Matthias Olbers and Franz von Paula Gruithuisen, who claimed that they had observed signs of civilized life on the Moon; and in like manner before Verne's stories of aerial journeys round the world there were the facts of the hot-air balloons, the hydrogen balloons, and the theories of the experimenters.

The countless possibilities of the age provided abundant material for every kind of projection; and a growing number of writers turned to the tale of the future as a means of exhortation, instruction or novel entertainment. Honoré de Balzac, the acclaimed master of realistic prose, tried his hand at several conjectural fantasies and in *La Comédie du diable* of 1831 set a man of the future to wondering about the ancients of the nineteenth century: 'What would they think, if they could see us travelling from here to St Petersburg in two hours, eating an ounce of gelatine which would contain enough nourishment for a year, preserving the old for an unlimited period in ice ... ?'[7] And in 1839 Edgar Allan Poe published the first of his prophetic stories, *The Conversation of Eiros and Charmion*, in which a passing comet puts an end to all life on Earth.

By 1842 the conventions of futuristic fiction had become so well established in England that Harrison Ainsworth thought it added to the attractions of the first number of *Ainsworth's Magazine* to include a prophetic piece, 'A Flight upon Flying'. This came from the young Martin Tupper who was then about to begin his notorious career as the most-quoted bad poet of the century. Again, in 1844 the editor of *Punch* felt so confident of his readers' interest that he introduced a nine-part serial on 'The History of the next French Revolution'. Thackeray was the author and the serial was a hilarious account of the time when a Bourbon would dispute the throne of France with His Imperial Highness Prince John Thomas Napoleon, a fourteenth cousin of the late Emperor.

From the evidence of Victorian technology Tennyson looked ahead to the conquest of the air; and in 'Locksley Hall' he saw '... the heavens fill with commerce, argosies of magic sails, / Pilots of the purple twilight, dropping down with costly bales.' The dream of aerial navigation was a popular fantasy of the nineteenth century. It appeared in Victor Hugo's *La Légende des siècles*, and it amused Hans Christian Andersen who wrote a brief article for the Copenhagen newspaper *Foedrelandet*, in which he described the time when passengers would traverse the Atlantic, 'flying across the ocean on wings of steam', to visit the ruins of Europe. The visitors would be Americans, of course, and their slogan – 'See Europe in eight days.'

In this way, decade by decade, the first men of the new industrial epoch evolved their own characteristic fantasies of time and space; and an examination of them will show how the growing certainty of coming changes had a decisive influence on the new themes in the tale of the future. The most instructive means of looking into this complex relationship between the intellectual climate of the period and the growth of futuristic fiction is to consider two remarkable books, *Le Roman de l'avenir* of 1834 and *Un Autre Monde* of 1844.

The author of *Le Roman de l'avenir* was Félix Bodin, a historian and member of the Chamber of Deputies, who won for himself the distinction of being the first man to write at length on the subject of futuristic fiction. In a two-page essay in *Le Miroir* in 1822, he invented the term of *la littérature futuriste* ('Like everyone else I have got the craze for making new words') and twelve years later he enlarged on his original ideas in *Le Roman de l'avenir*. This book, when taken in conjunction with the supporting evidence from the United Kingdom, shows that by the 1830s the 'tale of futurity' and *la littérature futuriste* had escaped from the constrictions of utopian fiction to become an accepted means of recording whatever anyone had to say – serious or humorous – about any aspect of the future. Every page of *Le Roman de l'avenir* shows that Félix Bodin expected his readers to be thoroughly familiar with the style and the matter of futuristic fiction. He began, for example, with a witty pastiche of Mercier. He imitated the preface to *L'An 2440*, but dedicated his book not, like Mercier, to the future, but to the past:

> Venerable Past, you have provided all the elements of this book, for when you had the advantage of being the present, you were already big with the future, as Leibnitz has put it very nicely. In dedicating this work to you, I am only returning what belongs to you – to use a phrase that does not belong to you in the least. You can see that I am not one of those inconsiderate people who are for ever looking towards the Eden or the El Dorado of future centuries, and are for ever heaping reproaches and insults on you, as though it had depended on you to have done better than was possible – you, poor victim sacrificed to the law of progress; you whom less fortunate generations have used as a stepping stone to the improvement and advancement of those who have followed them.[8]

Bodin then goes on to place the tale of the future in historical perspective by pointing out that 'in the past, when belief in the progressive degeneration of humanity was dominant, the imagination could only turn in fear to the future, painting it in sombre colours'. The pessimism of past ages has given place to the meliorism of the nineteenth century; and so, 'the future appears to the imagination in a blaze of light. Progress, understood as a law of life for mankind, has become at once a clear proof and a sacred manifestation of Providence.' The idea of progress carried the promise of continuing improvement for all men, and the tale of the future could draw upon the vast sources of the marvellous to describe what would be. Thus, 'in presenting the idea of perfectibility in an imaginative and dramatic narrative, the author of an epic story about the future finds a means of stirring and rousing the imagination, and of hastening the progress of humanity in a way far more effective than the best philosophical explanations presented with the greatest eloquence.'[9]

In spite of his enthusiasms, Bodin did not make the cardinal mistake of the forecasters, indeed also of the great Laplace: the failure to recognize that, whatever men can predict, they cannot predict the future state of knowledge. In the nature of things, Bodin asks, how is it possible to discover what the conditions of life in the twentieth century will be? Because men have failed to agree about so many things in the past, it is most unlikely that they should suddenly begin to agree about the kind of future they desire. Bodin had come upon the *ceteris paribus* factor that has confounded so many attempts at forecasting, for he had realized that unknown conditions and unexpected events can change the course of future history. He sums up these ideas in a striking passage:

> What political thinker in the ancient world was able to conceive of the possibility of a society without slaves? The great mind of Aristotle and the inspired thinking of Plato were unable – I will not say to have an inkling of but rather – to comprehend that a single town in Italy would end by conquering and civilizing a world twice as big as the one they knew. What genius in the seventeenth century could have formed any idea of what has been going on in two hemispheres for the last fifty years? That honest and at times amusing orator, Mercier, thought some fifty years ago of dreaming about the year 2440; and he did not even manage to think of representative government, trousers, and hair cut in the Titus style. He did not go much beyond the ideas of some French economists and philosophers who were in fashion in his day; and his account of a benevolent monarchy, which was only a modification of absolute power, appears hardly any more advanced than the hair-style of his future citizens for whom he reserved the bold innovation of making them whiten their hair with a touch of powder and tie it up in a chignon.[10]

After demolishing Mercier in this vigorous manner, Bodin started on his own vision of life in the second half of the twentieth century. What he foresaw was no great improvement on

Mercier; and what he had already said about the dominance of fashionable ideas in *L'An 2440* proved to be equally true of his own history of the future. During the twentieth century the Slav countries drive the Turks out of Europe, and Constantinople becomes a free city. In Palestine a company of Israelite bankers have re-established the kingdom of the Jews; in Italy there is a federal government; and in Ireland pigs and potatoes have helped to raise the population to a total of ten millions. Like Goethe, Saint-Simon and many others, Bodin was confident that the engineers would build a canal across the isthmus of Panama; and, in keeping with the favourite presumptions of his time, he assumed that industrial inventions, flying machines and advanced means of communication would lead to the ending of war, the abolition of slavery and polygamy, and the establishment of a world congress. The realities of political and social conditions in the Europe of 1834 dissolve into rosy images of universal happiness in the manner of the eye-catching patterns in Sir David Brewster's kaleidoscope, which was the plaything of the day. The twentieth century offers permanent absolution for all the failings of the past; and in that happier age 'the inter-marrying between races and the intermingling of the nations have gone on with increasing effect. The original differences between the various branches of the human race have been removed. The various languages have grown towards one another, and some of them have almost disappeared. The passions, beliefs and specific characteristics of the nations are now for the most part a matter of history; and the word *nationality* is beginning to have no more than a vague meaning.'

Although Bodin merits a commendation for the first popular study of the origins of futuristic fiction, the main value of his book lies in the evidence it provides for believing that the intellectual revolution implied in the idea of progress had been completed by the 1830s. The ideas of Condorcet, William Godwin, Robert Owen, Fourier and especially those of Saint-Simon had worked into the general thinking of Europe during the first quarter of the nineteenth century, and they had given a stiffening of theory to the belief in continuous change. There were, for example, Godwinian elements in Mary Shelley's *The Last Man*, and Bodin clearly drew on Saint-Simonian ideas about the reorganization of society. Political philosophies and fantasies of the future arose spontaneously from the ferment of the industrial revolution; for the new view of society – to take a title from Robert Owen – encouraged the growth of social schemes and of visual images that were an outward sign of the first major shifting within the imagination since the sixteenth century.

Whilst Fourier and Saint-Simon put forward their proposals for universal unity and the reorganization of the world, the poets and the painters traced the new shapes they perceived through the inner vision of the mind's-eye. The balloon, for example, acquired a symbolic function; it was a means by which the imagination was able for the first time to contemplate both the new-found capacities of mankind and the greatly enlarged scale of human activity. For writers of futuristic fiction the voyage through the skies was an opportunity to demonstrate the triumph of technology; but for the poet the bird's-eye view became a celebration of the variety of human life. Thomas Hood began a poem to Graham, the renowned balloonist of the 1820s, with a cry – 'Let us cast off the foolish ties / That bind us to the earth, and rise / And take a bird's-eye view!' And across the Channel the *vol d'oiseau* was a favourite device with Victor Hugo, especially in *Notre Dame de Paris* where he used it in the opening chapter in order to contain his vision of the unity of Paris.

These instances, however, are of slight interest in comparison with the originality and the extraordinary range of the topics in an illustrated book of 1844, *Un Autre Monde*, which is undoubtedly one of the most remarkable examples of the new literature generated by science in the nineteenth century. The author was Taxile Delord and the artist was Jean-Isidore Gérard, known as Grandville, the redoubtable caricaturist of the *Charivari* and the scourge of the July Monarchy. Their book is the case-history of imaginations that found delight in the innovations and marvels of the industrial revolution; but behind the verbal and visual fantasies author and artist kept constant watch for the inescapable contradictions they observed in human existence. They showed how mechanization had persuaded their contemporaries to regard machines as substitute human beings. The opening chapters describe a grand 'Mechanical-Metronomic Concert' at which all the instruments and the vocalists are operated by steam; and Delord notes in a most perceptive comment that 'in this century of progress the machine is a perfected man'. Then the text lifts off to take different sightings of industrial civilization, as Grandville in the person of a balloonist reveals the world in a new perspective from above.

Evidently the two men felt that they could only come at the truth by placing themselves outside their subjects – a carnival of animals at which the dancers wear human masks; puppets that display the mechanical conformities of social life; a nightmare vision of a revolt by the plants and trees which proclaim their new circumstances in the conventional language of the demagogues; a museum that presents in satiric fashion the art of the past, present and future; an aerial voyage to a surrealistic land where nature has consecrated the logic of social divisions by making midgets of the workers and giants of the upper classes; an ironic discussion with illustrations of the best form of government, which closes with an account of the labours of the phrenologists to make all men better and happier by the simple expedient of manipulating the heads of new-born children into the best psychological conformations.

The relaxed and conversational style of *Un Autre Monde* is proof enough that the new fiction had come of age. Like Félix Bodin, Delord expected that his readers would have no difficulty in accepting fantasies of other worlds, novel modes of life and extraordinary visions of future developments. Like his American contemporary, Edgar Allan Poe, Delord began from the possibilities suggested by the sciences; but there the resemblance ended, since a more genial temperament and a feeling for the grotesque diverted Delord from the sustained exploitation of the single themes that Poe developed in *Hans Pfaall*, *The Conversation of Eiros and Charmion*, *The Sphinx*, and *Mellonta Tauta*.

The real achievement of the book was not the text but the brilliant draughtsmanship of Grandville's illustrations. These speak directly to the modern reader of those changed conditions in the first industrial age that found their characteristic reflections in an original iconography of transformations, explorations and projections. For example, in the chapters devoted to 'Une Après-Midi au Jardin des Plantes' there are eight drawings of hybrid animals – winged frogs, an amphibious dromedary, a centaur, and the various species of the *Doublivores* that 'all have the distinction of being bicephalic, since they have a second head in place of that appendix commonly called the tail.' The tour of the zoological garden is, in effect, a commentary on the uses men might find for the sciences. Grandville makes his point with a public notice next to a collection of hybrid animals:

> THE HYBRIDS
>
> Ruminant birds, winged quadrupeds, pachydermatous insects: all these animals were born in the Jardin des Plantes. Thanks to the efforts of the Professor of Zoology, who developed these hybrids, they in their turn produce other crossed species which are as innumerable as those of the vegetable kingdom ... The *Pegasus* type, formed by crossing the giraffe, elephant or rhinoceros with the *Scarabaeids*, is in a special class called *Racers*, a name that does not need any etymological explanation. They learn easily, and they will prove of great advantage in transportation, since they can move rapidly and could even replace the railways.[11]

As the French were looking at the illustrations of *Un Autre Monde*, the British were recovering from the exciting news that someone had at last discovered the secret of powered flight. The story began with Sir George Cayley, the aviation pioneer, who helped to found the Adelaide Gallery in London as a centre for scientists and as a place where they could carry out experiments. There, on an unrecorded day in 1843, a civil engineer, W. S. Henson, brought a model steam-driven aeroplane which was propelled by two six-bladed pusher airscrews; and to the stupefaction of the observers the Aerial Steam Carriage trundled along the ground in a cloud of steam and took off into the air.

Henson rushed to the Patent Office and on March 28th filed the specifications of his proposed flying machine. On March 30th the first news appeared in *The Times*, and an excited nation learnt that an Aerial Transit Bill had been proposed to the House of Commons and that a group of promoters had formed an Aerial Transit Company. Then the lithographers followed with optimistic views of the *Ariel* in flight over St Paul's, the Thames, Hampstead Heath, Paris, Vienna, the Pyramids. By May, alas, after the engineering experts had begun to talk of the problems inherent in the power–weight ratio of the model – encouraging but not decisive trials, they said – the editors of the newspapers and the new illustrated magazines settled down to wait for the next report of some great advance in the conquest of the air.

This came in the following year and on the other side of the Atlantic. Again, it was the New York *Sun* which gave the unwary reader the astounding news: 'THE ATLANTIC CROSSED IN THREE DAYS! SIGNAL TRIUMPH OF MR MONCK MASON'S FLYING MACHINE!' And once again that enterprising publication reported what everyone wished to know, relating in the most circumstantial detail how 'Mr Monck Mason and Mr Robert Holland, the well-known aeronauts; Mr Harrison Ainsworth, author of *Jack Shepherd* etc., and Mr Henson, the projector of the late unsuccessful flying machine' had crossed the Atlantic by a new steering balloon in seventy-five hours. The report spread throughout the United States, and then came the realization that the journey of the Victoria balloon had been another hoax. At that time however, only the editor of the *Sun* knew that the author of the article was Edgar Allan Poe.

Poe had arrived in New York from Philadelphia the week before, with $4.50 in his pocket and a sick wife; and at once he put his imagination to work. Following his usual practice, Poe used known facts as the basis of his fabrication. He took over several descriptive passages in Monck Mason's account of his balloon voyage from London to Weilberg and he added some useful scientific detail from *Remarks on the Ellipsoidal Balloon.* The result was a work of art. The craft of the writer matched the ideas of the age, and in the emphatic opening sentences Poe expressed the fondest hopes of his readers:

> The great problem is at length solved! The air, as well as the earth and the ocean, has been subdued by science, and will become a common and convenient highway for mankind. *The Atlantic has been actually crossed in a Balloon!* and this too

> without difficulty – without any great apparent danger – with thorough control of the machine – and in the inconceivable brief period of seventy-five hours from shore to shore![12]

The modern reader has to forget the hoax in order to understand that the Balloon Hoax was part of, and depended on, the growing consensus on the future of mankind. The confident style of Poe's invention and the visual images of Grandville's *Un Autre Monde* show that the idea of the future had begun to find immediate and faithful expression in visions and projections of every kind. There is much more to be said about the varieties of futuristic fiction, and especially about the decline of the old-style terrestrial utopia and the establishment of the ideal state of the future as the natural form of literature for the technological and progressive nations. For the moment, however, it is enough to show, first, that the accepted image of the future was the product of an ever-widening discussion of anticipated changes and, second, that the supporting belief in a process of constant and beneficial development derived from the new histories of the nations and the new theories of organic and social evolution.

The weekly illustrated magazines had a particularly important role in spreading information about the latest advances in the sciences and in encouraging their middle-class readers to speculate about the most probable course of mechanical invention in balloons, railways, steamships, ships of war, artillery, submarines and the Channel Tunnel. An examination of the *Illustrated London News* or *L'Illustration* will show that it was editorial policy throughout the last century to give regular reports on recent progress in the applied sciences. These were often one-page, fully illustrated accounts which were designed to suggest by word and image the exact shape that some project or invention would assume in the future. Typical examples of this popular activity can be seen in the *Illustrated London News* for October 11th, 1845, which carried a full page of text and illustrations on the proposals for 'Samuda's Atmospheric Railway'; and in the detailed discussion of 'The Proposed Anglo-Gallic Submarine Railway' on November 10th, 1855. Another full-page piece on 'The Mail Coaches' and 'The

Arriving Train' made the standard connexion between past, present and future:

> Snake-like it comes exulting in its strength,
> The pride of art – the paragon of skill.
> Triumph of mind! what hand thy bound shall mark?
> Seen looming palpably 'mid cloud and dark,
> Yet other triumphs, more than this sublime,
> Rise numerous on the far-seeing ken
> Of those who watch, and hope the good of men.[13]

At the same time, as the inventions and the projects increased in number, there was a corresponding growth in the histories of science that had begun to appear during the first quarter of the nineteenth century; and in the 1860s their authors moved into the expanding market for juvenile literature by providing admirably illustrated books with such fitting titles as *The Romance of Invention* and *The Marvels of Science*. They were offerings on the altar of a victorious materialism, and they were generally presented in a language of great objectives, heroic strivings, unprecedented advances and immense blessings. All of them expressed views also to be found in *The Communist Manifesto*, for their authors held with Marx that the achievements of the nineteenth-century bourgeoisie were a crucial stage in the progress of mankind. Although few of them could have agreed with Marx's theory of the predestined advance to the classless society, they all joined in his whole-hearted celebration of a universal and beneficent materialism: 'Subjection of nature's forces to man, machinery, application of chemistry to industry and agriculture, steam navigation, railways, electric telegraphs, clearing of whole continents for cultivation, canalization of rivers, whole populations conjured out of the ground – what earlier century had even a presentiment that such productive forces slumbered in the lap of social labour?'[14]

Whilst Marx laboured through two decades to enshrine his presentiments about the destiny of mankind in *Das Kapital*, his contemporaries set down their own visions of the future in histories of progress, accounts of technological developments and predictions about coming events. In 1852, for example, Michael Angelo Garvey traced the rainbow curve of progress from the contemporary conquest of nature to the greater victory over human nature that would follow on the ending of war. His book was *The Silent Revolution; or the future effects of Steam and Electricity upon the Condition of Mankind.* In the opening pages he discoursed on the happy union between science and society:

> No one can contemplate the unexampled progress of science within the present century, without feeling that a new epoch has commenced in the history of our race. The divine powers of the mind are extending their grasp and rising to a state of higher activity. Fields of knowledge undreamt of in the earlier ages of the world are successfully cultivated. The farthest regions of space are explored and the secrets of their starry depths unfolded to men.[15]

This and much of what followed was part of the popular theory about progress; but later on, when the author describes the changes he expects, it is possible to feel the full power of the Victorian belief that the sciences would transform the world:

> The capacity of all soils will be tested, and nature herself assisted in bringing her richest productions to a maturity hitherto unattained. This process will not be stayed by the limits of man's present domination over the earth. It will enlarge his domain; our race will be more equally distributed throughout the world. When the agencies of transport are universal, men will no longer consent to be crowded together in towns and cities, whilst the pure air and joyous landscape are easily accessible, nor to jostle one another in millions within the confines of narrow islands and territories, snatching the bread from one another's mouths, whilst in other lands, and beneath skies as serene, the rich bounty of nature ripens and decays untouched by man. The fruitful continent of Australia, sufficient in itself to sustain the whole population of Europe, will in time be traversed with railroads. The vast though hitherto impenetrable regions of Africa will be opened by the same mighty agency. The broad Savannahs and measureless Steppes of America will be brought within the scope of human industry, and the wild-horse and buffalo which now roam over them in

> undisturbed possession be yoked to the plough. The half-peopled regions of the north and east, the beautiful but desolate islands of the Indian Sea, will be made accessible to the enterprise of mankind; an amount of territory fifty times greater than that man now inhabits and cultivates still lies unwrought and unpossessed upon the surface of the globe; it will one day furnish its full tribute to the necessities of our species.[16]

That phrase about the territories 'unpossessed upon the surface of the globe' has the unintended effect of telling the twentieth-century reader that the happy prospect of an expanding population was not a programme for the future of all mankind. Buried beneath the planner's prose (those casual phrases about other lands and half-peopled regions) was the imperial ambition to secure and to develop the resources of the world in the interests of the new technological societies. In a contradictory and self-deceiving way, very typical of so many nineteenth-century forecasts, the author assumed that desirable social changes would be the first fruits of material progress. He expected that, as the steamship and the railway opened up the world for the Europeans, the new means of communication would help 'to draw all nations into more intimate connexion and to convert the whole human race into one society'. Had Michael Angelo Garvey considered the more intimate relationship that sprang up between the British and the Chinese during the Opium War? Had he considered the events in India during the decade before his book came out – the First Afghan War, the Sind War, the Sikh War, the Annexation of Sattara, the Second Sikh War, the Annexation of the Punjab, the Second Burma War?

This unfortunate discrepancy between real events in the nineteenth century and the imagined paradise of the future reveals the mythic power that the idea of progress had acquired. For the unsuspecting it was a Faustian contract between science and society – the promise that men would be able to do whatever they wished with nature and that they could even bring about lasting peace on earth:

> The hour is coming, hastening with the momentum of ages, when the 'ultima ratio regum' – the argument of murder,

> rapine, famine and pestilence – shall be banished from amongst men, consigned to the chamber of horrors, in which history preserves the memorials of crime ... The railway signal is the knell of war, the anticipative requiem of that military thing so monstrously miscalled glory ... A new light is dawning upon the world which will render such monstrous deceptions impossible; men, by a more perfect association with one another, and by the more perfect blending of their real interest which must ensue, will see in all the broadness and perspicuity of truth that the clumsiest and least rational way of disposing of enemies is to kill them.[17]

These ideas about the power of science appeared on all sides. They were a source of many utopias of the future and they were part of the new histories then being written throughout Europe. In 1837, when Macaulay was at work in India on the codification of the criminal law, he wrote a long review of a new edition of Bacon's works; and in his aphoristic and exclamatory way he announced that Utility and Progress were the signs of the age, and that Bacon was the true begetter of nineteenth-century industrialism. Macaulay found (so he thought) that before Bacon appeared philosophy had begun and ended with words, whereas 'the philosophy of Bacon began in observations and ended in arts'. Ask any follower of Bacon what the new philosophy had done for mankind, and the answer would be that it had changed the condition of life: 'It is a philosophy which never rests, which has never attained, which is never perfect. Its law is progress. A point which yesterday was invisible is its goal today, and will be its starting-post tomorrow.'[18] This was the language of the Poet Laureate in 'Locksley Hall' and of the Prince Consort before the opening of the Great Exhibition.

This thinking was the source from which Jules Verne drew his ideas for dreams of future achievements and it was central to the philosophy of social progress that Henry Thomas Buckle presented in his *History of Civilization in England*. Buckle had spent fourteen years in investigating the effects of environment on the evolution of society; and in 1857 he explained in his first volume how favourable circumstances had set the peoples of Europe on the road to prosperity. 'And it is accordingly in

Europe alone,' he argued, 'that man has really succeeded in taming the energies of nature, bending them to his own will, turning them aside from their ordinary course, and compelling them to minister to his happiness, and subserve the general purpose of human life.'[19] Buckle held that, because of the universal and uniform operation of the progressive principle in society, it was possible to follow the sequence of developments from their origins in the past to their conclusion in the foreseeable future. In a footnote on Montesquieu and Turgot he revealed in passing that he 'believed in the possibility of generalizing the past so as to predict the future'. Montesquieu would have reached that conclusion, he thought, 'had he lived in a later period and thus had the means of employing in their full extent the resources of political economy and physical science ... As it was, he failed in conceiving what is the final object of every scientific enquiry, namely, the power of foretelling the future.'[20]

Buckle had composed his history according to the canons of the prevailing positivism and historicism of his time. The glorious sequence of progress and improvement, it seemed, had reached a climax in the Victorian period, which was the best of all historical periods, just as European civilization, then securing its hold on all the continents, was the best of all civilizations. Like many of his contemporaries – Herbert Spencer, William Lecky, Michelet, Macaulay – Buckle had written under the influence of an illusion, the Promethean persuasion that man had become the master of the universe. The new creed had a numerous and most influential following throughout Europe and the United States; and all who accepted the doctrine of progress made their confessions of faith in a future paradise of continuous improvement. In 1858 the Honourable George Bancroft, late U.S. Minister in London and a most industrious exponent of Herbert Spencer's philosophy, spoke to the New York Historical Society on 'The Progress of Mankind'. He urged his audience to be steadfast in their labours for the better world-to-come: 'Since the progress of the race appears to be the great purpose of Providence, it becomes us all to venerate the future. We must be ready to sacrifice ourselves for our successors as they in their turn must live for their posterity ... Everything is in movement, and for the better, except only the fixed eternal law by which the necessity of change is established.'[21] The American orator had no means of knowing that Darwin was at work on the final draft of the *Origin of Species*, which was to be both Genesis and Deuteronomy in the Victorian book of progress. When the *Origin* appeared in 1859, it explained the fixed eternal laws of nature as a delicate balance between life and death; and Darwin was accepted as the Moses of a new revelation who brought together all the earlier intuitions and ideas about evolution and progress in his unique theory of natural selection.

The new urban societies were ready for the Darwinian proposition that, 'as natural selection works solely by and for the good of each being, all corporeal and mental endowments will tend to progress towards perfection'. Everyone had seen how medical science and the new technologies had acted on the condition of human life in ways that paralleled the operations of nature, as Darwin described them in the concluding paragraphs of the *Origin of Species*: 'Whilst this planet has gone cycling on according to the fixed law of gravity, from so simple a beginning endless forms most beautiful and most wonderful have been, and are being, evolved.'[22] The language of the zoologist proclaimed the incontrovertible experience of progress. When Darwin rounded off his final remarks about the evolution of organic life by hazarding 'a prophetic glance into futurity', he presented his theories in the manner and the idiom of contemporary practice in forecasting and in futuristic fiction. The universal sense of then-and-now, which had grown out of so many changes and developments, encouraged all to see all things as having their special part in the onward movement of mankind. Darwin, for instance, found good news for his readers in the evidence of the past. 'As all the living forms of life are the lineal descendants of those which lived long before the Silurian epoch', it followed that there would be a steady development of life and that everyone could 'look with some confidence to a secure future of equally inappreciable length'.[23]

This argument from the evidence of things past to the shape of things-to-come was the driving force in most of the contemporary stories about the future, and it was a favourite source of instruction for the young. In 1859, for example, whilst some of the adults were reading their Darwin, the children had the opportunity to read the latest history of invention, *The Triumphs of Steam*. The lesson began in this way:

> 'Only fancy, Aunt Helen,' added Charles, 'that Uncle Harry was in Paris yesterday and will be at home today. Is it not wonderful? What did people do before there were railroads and steam-boats?'
>
> 'Went by coach or van, or in their own carriages, or on foot as the case might be.'
>
> 'What a waste of time,' exclaimed Charles. 'It was a good thing that steam was discovered.'[24]

By the 1860s, then, the philosophies of history and the theories of evolution had merged with the general conviction of universal progress to establish a paradigm of existence. The idea of progress was a law of being and becoming. It explained itself with total clarity to every level of understanding. For the young it was a good thing that steam had been discovered, and for their parents it inspired self-congratulatory ditties about the miracles of human inventiveness:

> What would our Fathers say, if from the dead
> They, for a time, could raise their ancient head?
> What would they say if they could only see
> The great results of Electricity?
> How would they stare to see our measured light,
> Our cities lit with gaseous lamps at night?
> What would they think of all-puissant steam
> Treading the Earth, the ocean, lake and stream?
> What would they think of locomotive speed
> Uniting distant lands of every creed?[25]

These complacent ideas about the progress of the age were the popular expression of most powerful beliefs. These beliefs found a more ideal form in utopias of the future, and they stated their essential theory in the new philosophies of progress that began to appear from the 1850s onwards. Once again the best evidence of this advance in the idea of the future comes from France. The increasing sophistication in the theory of progress can be noted in the difference between the first cautious propositions of Turgot in the *Philosophical Review* of 1750 and the earliest exhaustive analysis of the established doctrine in Javary's *De l'Idée de progrès* of 1851. By that date the accumulated evidence of one hundred years was sufficient for Javary to write at length on the various kinds of progress, on the question of moral progress, and on the two principal groups of writers interested in the idea of progress – 'those dominated by the historical point of view or by the analysis of past advances, and those dominated by the utopian vision or by the search for the ideal state towards which mankind should direct all its efforts.'

In contrast to the French practice of systematizing the experience of the human race, the British mode of explanation favoured a native pragmatism and a logical argument from the evidence of technological advances. The titles speak for themselves: G. R. Porter, *The Progress of the Nation*, 1836; R. K. Philp, *The History of Progress in Great Britain*, 1859; N. Arnott, *A Survey of Human Progress from the Savage State to the Highest Civilization yet attained*, 1861. After that, in keeping with the general movement towards a more detailed examination of future probabilities, the histories of progress developed into rudimentary studies of foreseeable advances in society and in human conditions. Thus, Charles Dollfuss started his investigations in *Le Dix-neuvième Siècle* in 1865 by claiming that 'it is already possible to divine the shape of future things by virtue of the laws that, without clamour and within the depths of the human spirit, forge the indestructible chain of progress ... Our future is the complete transformation of society by science and by liberty.' A similar study appeared in English in 1866, *Thoughts on the Future of the Human Race*, in which William Ellis examined the factors that would enable an investigator to discover the shape of coming things. That this had become common practice by the 1860s is evident in the author's assumptions:

> Students of the present day, striving to acquire a knowledge of the future, should not be forgetful of the advantages which they enjoy, as compared with their predecessors. Never before was there so large a number of prognostications and

> phenomena on record ready at hand for examination. The stream of time, which has been continuously passing the future into the present and past, has shown how far actual events and phenomena have corresponded with those prognosticated.[26]

This was a familiar concept of the age. It was central to all that Herbert Spencer wrote. It controlled the thinking in Henri de Ferron's *La Théorie du progrès* of 1867 and of Edmond About in *Le Progrès* of 1864. All these philosophers and forecasters insisted on the certainty of continuing progress, and they repeated About's theorem that '... there is not a single intelligent man who does not feel himself joined by invisible ties to all men – past, present, and future. We are the heirs of all who are dead, the associates of all who are living, the providers for all who will be born ... We are better and happier than those who have gone before us, and so we should see to it that our successors will be better and happier than us.'[27]

Since there was an eternal bond between all things, it was customary to recite the paradigm of progress from past-historic to future-perfect. The sequence from past to present, for example, appeared in the account of the ending of the Ice Age in Switzerland which Karl Vogt, the Professor of Natural History in the University of Geneva, gave in his *Lectures on Man; his place in Creation and in the History of the Earth*: 'The sea rose, the land became warmer, the ice-shroud melted, the highest ridges showed their pinnacles, the ice broke into separate glaciers, which continued to fill up the valleys ... This prehistoric history is no romance; it is derived from actual facts and the deductions therefrom.'[28] From the facts of the past there followed deductions about the future; and in his account of *Man in the Past, Present and Future*, the German scientific writer Professor Ludwig Büchner traced the progress of mankind from the Aurignacian Era to the paradise of the coming centuries. His book, which ran to many editions and was translated into five languages, set out to give 'a popular account of recent scientific research as regards the origin, position and prospects of the human race.' Büchner had much to say about the blessedness of mankind. He knew that the guiding principles of progress and evolution could not fail to make the world better and better. 'In other words,' he prophesied,

> the struggle for the means of existence will be replaced by the struggle for existence, man by humanity at large, mutual conflict by universal harmony, personal misfortune by general happiness and general hatred by universal love. With every step in this path man will depart more and more widely from his past animal condition, from his subjugation to the forces of nature and their inexorable laws, and approach more and more to the ideal of human development. On this course he will again find that Paradise, the ideal of which floated before the fancy of the most ancient nations and which, according to tradition, was lost by the sin of the first man. The only difference will be that this Paradise of the future will not be imaginary but real, that it will come not at the beginning but at the close of our development, and that it will not be the gift of the Deity, but the result of the labours and merits of man and of the human intellect.[29]

The professor had removed the Deity from the system of the universe, because the idea of progress and the theory of organic evolution had provided more than sufficient reason for a secular religion of human creativeness and continuous improvement. Even the more cautious Huxley was confident that '... thoughtful men, once escaped from the blinding influences of traditional prejudice, will find in the lowly stock whence man has sprung, the best evidence of the splendour of his capacities; and will discern in his long progress through the Past a reasonable ground of faith in his attainment of a nobler Future.'[30]

This was the language of expectation and these were some of the dominant ideas of the age that the young found in their books. During the 1860s a new trade in illustrated gift-books about the past and about the great inventions of the century had started in Europe; and these gave their young readers their first lessons in the ideology of progress. They taught through excellent engravings and by placing the subject – dinosaurs or railway engines – at the appropriate stage in the onward movement of the human race; and from the explanation of the past it was customary to look ahead to developments in the future. In 1863, for instance, the very successful and gifted popularizer of science Louis Figuier produced an illustrated account of geological history and evolutionary ideas in *La Terre avant la déluge*; and his book, like those of his contemporary Jules Verne, went into many editions and translations. In his last chapter Figuier turned to the future: 'Having considered the past history of the globe, we may be permitted to give a glance at the future that awaits us ... '

Another popular publication of 1865, *L'Homme depuis cinq mille ans* by Samuel Henri Berthoud, made the same linkage between past and future. The text gave an imaginative reconstruction of the principal stages in the evolution of human society, beginning with the first inhabitants of the Paris region and an illustration of mastodons roaming the Montmartre area in the Quaternary and ending with a chapter on the inventions of the year 2865. In this way the expectations of the parents passed on to the children and the idea of progress became part of the wisdom of the race. In 1865, for example, Charles Knight looked back over the forty years of his most effective efforts for the Society for Diffusing Useful Knowledge, and in his autobiography summed up the experiences of his life in the belief that 'Man is achieving a victory over time and space of which the imperfect beginning called forth our wonder, but we scarcely know how to contemplate the possible end without something like awe.'[31] Those were the themes and almost the exact words of another illustrated book of 1865, *A History of Wonderful Inventions*, which begins with a dominie's declaration in favour of progress:

> The triumphs of science at present realized may seem but trifles in the future. The human mind enlarges with its conquests, and each new step gives us encouragement to proceed with another, and we see not where a limit can be placed to the grand dominion man may in the end obtain. A period may arrive when even the Steam-engine may be derided as an imperfect piece of mechanism, and some discovery made that will enable man to wield equal force without the employment of its cumbrous bulk and expensive fabrication.[32]

Over the page the author puts the Victorian young persons in their place by indicating what should be their role in adult life. 'The glory of the future', the teacher says, 'is only to be realized by maturing the grandeur of the present. It is by going on from the point already attained that a more splendid future is to be reached.' These are the ideas but not the words of Buckle and Macaulay. As the author calls on the young to join in the glorious work of progress, he retails the central beliefs of the age in the hackneyed phrases of improving literature – service to knowledge, gird up his loins, devoted and untiring disciple, students are what science requires, benefactor of mankind, banner of science, cultivation of fraternal good-will, inevitable tendency to bring war to an end. His promise is that science will change the world for the better. 'Who, then, would be slack to enrol himself her disciple while life is young, and there is a prospect of blessing mankind by entering her service ... Young reader, the glorious path is open; none can prevent your entering it; there needs but patience and resolution.'[33]

4

From Jules Verne to H. G. Wells. The Facts and the Fiction of Science

By the 1860s the various progressive and evolutionary theories of the nineteenth century had coalesced to form a pattern of general expectations. This was the idea of progress, and at its most simple it was no more than the universal agreement that things had changed for the better and would go on changing. In its more vulgar and materialistic form this appeared as the frequently proclaimed belief in technological advances and in political developments that would increase the well-being of the great industrial societies. Between these two extremes all the varieties of thinking and believing found their specific modes in theories of progress, philosophies of existence, forecasts of coming changes, and in the consolidation of the tale of the future.

The many ideas about the past and present – especially the social and political theories – worked themselves into declarations that were at once dogmatic and systematic. These discovered the seeds of future societies in history and in contemporary events. They presented conclusions about the direction of social advances that varied according to individual temperament and the interpretations of the evidence. It would be the best of worlds; and sometimes it would be the worst of worlds. The major system-builders and propagandists – Marx, Engels, Spencer, Nietzsche, Treitschke – taught that conditions in their time would lead to the victory of the proletariat, or a perfect industrial world, or to the coming of the new race of Overmen, or to the great war for German colonies. At the same time, from the 1860s onwards, there was a very rapid increase in the numbers of ideal states of the future, prophecies of coming wars and scientific romances. By 1871 the tale of the future had become the most popular and sometimes the most effective means of examining, describing and prescribing whatever writers thought would happen, or feared could happen, or considered should happen in the years ahead. The special conditions responsible for the immediate success of *L'An 2440* in 1771 had become part of the norm for the technological nations of 1871; for it seems that the growing uniformity of life in the great industrial communities required an imaginative correlative that would give shape to the hopes, fears and fantasies of urban man. Only the tale of the future could meet these rigorous specifications, and so it came about that by 1871 nation began to speak to nation in the new literature of visions and prophecies.

The scale of operations had changed both in fact and in fiction, and the extent of that change is evident in the difference between the European response to the eighteenth-century vision of *L'An 2440* and the world-wide reception of so many stories about the future during the last thirty years of the nineteenth century. The improved facilities for communication in the printing telegraph and the rotary press coincided with the beginnings of universal literacy; and the abundant evidence of word and image caused all to read the signs of future events in the sinking of the *Merrimac* in 1862, in the decisive role of the breech-loading rifle in the Austro-Prussian War of 1866, in the change in maritime and imperial communications that would follow on the opening of the Suez Canal in 1869, and most of all in the consequences for Europe of the German victories in the war of 1870. On July 1st, 1865, for example, when all still seemed to be going well with the laying of the Atlantic cable, the editor of the *Illustrated London News* looked into the future and told his readers that the new electric telegraph would affect everyone throughout the world: 'For instantaneous communication between America and Europe means, of course, in its ultimate development, instantaneous communications all the world over. And so we shall have daily before our eyes a bird's-eye view of human affairs over the entire surface of the globe, and we shall be able to study all nations, as day by day they are making contemporaneous pages of history.'

Twelve months after that forecast the prophetic events in a serial story of lunar travel had engaged the enthusiastic attention of the readers of the *Journal des débats*. The success of *From the Earth to the Moon* launched Jules Verne on his long and very profitable career as the first reigning pontiff in the history of science fiction, since the interest his Parisian readers found in the ingenious details of his space journey confirmed the reputation he had won for himself with his first story, *Five Weeks in a Balloon*, in 1863. Success had not come easily to Verne, who had left Nantes in 1848 to seek his fortune in Paris. He tried his hand at the law, at writing comedies and operettas, occasional journalism and that last gamble of the hopeful – the Bourse. In 1851 he began to talk of writing a *roman de la science*; he took to studying scientific subjects at the Bibliothèque Nationale and he wrote occasional pieces for the *Musée des Familles*. And then the golden legend of Verne's life as a writer began in the October of 1862. He brought a manuscript to the publisher P. J. Hetzel, who read the draft and gave Verne the good news: tighten it up, make it into a complete story, and bring it back as soon as possible. In fifteen days Hetzel had the revised version of *Five Weeks in a Balloon*, and a famous collaboration had begun. The fabulous story goes on to relate that Verne hastened to say goodbye to his friends at the Bourse: '*Mes enfants*, I am going to leave you ... I have just written a new kind of novel, all my own. If it succeeds, it will be the vein that leads to the gold mine.'

Verne soon discovered he could exploit the richest mine in the popular fiction of the nineteenth century, thanks to his gift for absorbing the ideas of others and to the enterprise of his publisher. Hetzel was the most original and successful French publisher of his time; and for some months before he met Verne in 1862 he had been thinking of a venture into the expanding market of educational literature for the young. There was room, he thought, for a juvenile magazine of quality which would have to be well illustrated and would present the achievements of the sciences in an interesting manner, probably in the form of fiction. And then Jules Verne came into his office with his manuscript, and on December 24th, 1862, Hetzel published *Five Weeks in a Balloon*, the first of the forty-seven *Voyages extraordinaires* which Verne was to write during the remaining forty-three years of his life. The immediate success of the story convinced Hetzel that he had found the ideal writer for his projected magazine. The two men entered into a contract – Verne was to provide not less than one story a year for the *Magasin d'éducation et de recréation*, and each story would be republished as an illustrated book in the series of the *Voyages extraordinaires* which Hetzel started in 1867 with *The Adventures of Captain Hatteras*. In his introduction to that story Hetzel described the plan of the series. It turned out to be the production programme for the rest of Verne's working life:

> As the new works of M. Verne appear, they will be added to this series which we will always be careful to keep up to date. The stories already published and those still to come will together fulfil the plan the author had in mind when he gave his work the subtitle of *Voyages dans les mondes connus et inconnus*. His intention is, in fact, to summarize all the knowledge of geography, geology, physics and astronomy which modern science has brought together, and in that attractive style of his to rewrite the story of the universe.

Events soon proved that Hetzel was right in his claim that Verne had appeared at the right moment. After his first successes with the young he became a favourite with the readers of the most respectable newspapers. The sales of the *Journal des débats* were trebled when it serialized *From the Earth to the Moon* in 1866; and when *Le Temps* began the serialization of *Round the World in Eighty Days*, the world-wide interest was so great that British and American correspondents cabled the latest information on the exploits of Phileas Fogg to their newspapers. As the stories followed year by year, the profits and the glory were multiplied. The French Academy commended his works; the stage version of *Round the World in Eighty Days* became the rage of Paris; Pope Leo XIII received Verne in private audience and spoke approvingly of the wholesome quality of his stories; and there were all those translations into the principal languages of the world, including the editions in Arabic, Turkish, Chinese, and Japanese.

The inauguration of the modern period in the course of futuristic fiction coincides with the beginning of the modern world about 1870. Ever since then a distinguishing feature of the genre has been a singular capacity to transcend all national and cultural frontiers, so that from time to time the tale of the future has been able to achieve a degree of instant success never known before in the history of fiction. The earliest evidence for this can be seen in the international interest in Jules Verne's *Twenty Thousand Leagues under the Sea* in 1870 and in the world-wide commotion that followed on the publication of the *Battle of Dorking* in the May issue of *Blackwood's Magazine* in 1871. The author, Sir George Tomkyns Chesney, had written the first truly effective account of a future war and at once he found millions of readers throughout the world. The overseas editions in New York, Philadelphia, Toronto, Melbourne, and Dunedin plus the translations into Dutch, French, German, Italian, Portuguese, Spanish, and Swedish plus the seventeen counter-attacks and imitations in English – these offer compelling evidence of the way in which an imaginative projection, derived from factors common to all large industrial nations, could in the space of three months arouse the interest of innumerable readers for whom it had never been intended.

The instant notoriety of the *Battle of Dorking* and the rapid success of Verne's early stories were unmistakable signs of the continental change that had taken place since the transatlantic diffusion of the Lunar Hoax in 1836; Verne and Chesney scored the first world success in the history of the new communications industry. Two original writers had responded to the peculiar conditions of their time: the one by adapting the old-style imaginary voyage to the novel circumstances of urban society and technological progress; the other by converting the tale of the war-to-come into a popular and deliberately political device for demonstrating the dangers that seemed to threaten an island nation in the new epoch of breech-loading artillery and iron-clad ships of war. Between them they mark off the extremes of the tale of the future – peace and war, terror and delight, the perils of the nation and the mysteries of the universe.

Again, the works of Verne and Chesney present the first classic examples of specific functions in the tale of the future. Chesney gives a faultless demonstration of the tale of the war-to-come. He shows that the secrets of the game are a fluent narrative, realistic detail and a calculated bias. In a comparable way the wonders of science displayed in *From the Earth to the Moon* and in *Twenty Thousand Leagues under the Sea* are the first major examples of the way in which the conventions of science fiction allow a writer to devise fantasies about the titanic powers of modern man. The long and unbroken success of the *Voyages extraordinaires* is particularly instructive, since their world-wide dissemination owed much to the commercial enterprise of the Hetzel company. For the first time in the history of fiction the much increased capacities of the publishing trade allowed an editor to plan production from the first appearance of a story in a magazine to serialization in the daily press, and then to the special issues and the authorized translations into foreign languages and the long-term contracts with overseas publishers.

The careful control of the end-product began with the writing of Verne's stories. The correspondence between Verne and his publisher shows that the two men invariably discussed the work in hand and that Verne was often willing to accept the suggestions of Hetzel. He cut down the dialogue in his first story; he removed a duelling scene from *The Adventures of Captain Hatteras*; and he even agreed that Captain Nemo should not appear as an exiled Polish patriot for fear of reducing sales in Russia. Was it Hetzel who suggested the practices that are part of the basic formula in the Verne stories? The pious Catholic has surprisingly little time for the clergy, and most of his references to religion are so general that one suspects Verne of working for the neutrality that does not offend. None the less, the French patriot writes in the accepted style of European nationalism. At every opportunity he castigates the British for their colonialism. He is even more ferocious with the Germans, whose beastliness he portrays in the diabolical scheming of Professor Schultze, the demon scientist of *The Begum's Fortune*; but for his own countrymen he reserves discreet praise and some of the best roles. There is the always resourceful Hector Servadac, the good Dr Sarrasin who is the virtuous counter to the wicked Schultze, and there is Michel Ardan who in *From the Earth to the Moon* shows the would-be American astronauts what they should do in a chapter appropriately titled 'How a Frenchman Manages an Affair'.

This evidence of apparent calculation in Verne suggests a paradoxical reason for the world-wide interest in his stories. Although Verne was undoubtedly a great innovator, his success turned on the fact that he lacked original ideas. He owed everything to the enterprise of his publisher, to a gift for converting the facts of science into the stuff of fiction, and most of all to the way in which his fast-moving narratives exploited the darling ambitions of the age by developing themes that had been constants in the popular literature of the nineteenth century. The balloon adventures described in his first story had numerous antecedents in Julius von Voss, Cousin de Grainville, Mary Shelley, Jane Webb, and especially in Edgar Allan Poe, from whom Verne borrowed on many occasions. In 1857, for instance, a long-forgotten writer by the name of Saint-Elme Bernard published a balloon story, *Voyage aérien de Batavia à Marseilles*, in which he anticipated the historical explanations, scientific details and adventurous episodes that were the primary constituents of *Five Weeks in a Balloon*. In fact, there is a suspicious correspondence between the eye-witness accounts of Africa as seen from above in the Verne story and the aerial observations made in the earlier balloon journey from Batavia across the Pacific and the United States.

Again, the subterranean adventures in Verne's second story, *Journey to the Centre of the Earth*, followed in a well-established literary tradition that began with the French imaginary voyages of the seventeenth century and became one of the most popular topics in European fiction after the appearance of Baron Holberg's *Journey to the World Underground* in 1741. By the early nineteenth century the notion of a hollow earth had spread to the United States, and in 1818 Captain John Symmes sent a memorandum to the governments and the principal institutions of all major countries with proposals for an expedition to the centre of the world: 'I declare the earth is hollow and habitable within ... I ask one hundred brave companions, well equipped to start from Siberia in the fall season with reindeers and sleighs.' Ten years later the hollow earth theory appeared once again in *Die Unterwelt*, in which a German author set down the reasons – mythological, historical, geological – for thinking that there must be caverns measureless to man deep down within the Earth. And five years after that, in 1833, the idea appeared as an element in *MS Found in a Bottle*, the story that started Edgar Allan Poe on his career as a writer. Verne read that story with the close attention of an apprentice writer in search of a satisfactory style; and he learnt much from Poe's ideas and methods. He had, for example, the benefit of Poe's narrative technique in the Balloon Hoax of 1844 when he wrote *Five Weeks in a Balloon*, and there is clear evidence of borrowings from *Hans Pfaall* in Verne's third story, *From the Earth to the Moon*. Indeed, that story and the handling of the archetypal journey to the underworld illustrate Verne's gift for finding his themes within the substrata of literary tradition and the subconscious.

The voyage to the Moon is one of the most ancient myths. The literary history of the lunar journey begins in antiquity with Lucian's *True History*, and it advances through Kepler's *Somnium* into the numerous accounts of planetary journeys in the seventeenth and eighteenth centuries. In the nineteenth century there were various precursors of Verne; and their stories show a steady movement away from the dreams and fantastical excursions of the earlier Moon journeys to the scientific descriptions that underpin the narration in *From the Earth to the Moon*. In the United States the new means of propulsion for space vehicles was an improved application of the anti-gravity device that had once been the source of power in the flying island of Laputa. It appeared for the first time in George Tucker's *Voyage to the Moon* of 1827 and again in J. L. Riddell's account of *Orrin Lindsay's Plan of Aerial Navigation* in 1847. In French fiction the astronauts travelled through space thanks to an anti-gravitational device in *Star ou ψ de Cassiopée* of 1854; or by balloon in *Aventures d'un aéronaute parisien dans les mondes inconnus* of 1858; or by the first rocket motor in nineteenth-century fiction, which began the space journey in Achille Eyraud's *Voyage à Vénus* in 1865. This international advance towards a more realistic description of the journey through space is confirmed by the scientific details given in an anonymous English story, *A Voyage to the Moon*, which came out in a French translation in 1864 when Jules Verne was working on his own space story.

Within three months of the publication of *From the Earth to the Moon* American publishers were negotiating with Hetzel for the translation rights, and within five years Jules Verne had become one of the best known of French writers. His success with juvenile readers does not call for any explanation; but the great attraction many of his stories had for adult readers raises interesting questions about the social role of science fiction in the last century. The serialization in important newspapers of those stories that dealt with the anticipated triumphs of technology – *From the Earth to the Moon*, *Round the World in Eighty Days*, *The Clipper of the Clouds* – showed that what was good reading for the young had a revelatory quality for the adult. Verne was, in effect, composing idealized visions of human achievement, which he presented in terms of nineteenth-century experience as the results of human will and scientific knowledge. The scientist is at the centre of these stories in the same way as the capacity of the sciences was the core of the contemporary belief in progress; and so it will be no surprise to find that the possibilities Verne developed in his stories were part of the continuing discussion of the age. For example, one of the more widely read books of the 1870s was *The Martyrdom of Man*, in which Winwood Reade anticipated the central idea in *The Clipper of the Clouds* with his forecast of the inventions still to come: 'The first is the discovery of a motive force which will take the place of steam, with its cumbrous fuel of oil or coal; secondly, the invention of aerial locomotion which will transport labour at a trifling cost of money and time to any part of the planet.'[1]

This feeling of total self-confidence was a major source of power in the Verne stories. One of the secrets of his predictive fiction was the skill with which he carried forward the aspirations of the age to their triumphant realization in ocean-going submarines and vast flying machines. For instance, Winwood Reade had written of the time when, 'the earth being small, mankind will migrate into space, and will cross the airless

Saharas which separate planet from planet, and sun from sun'.[2] Those were the thoughts of Michel Ardan, the debonair young French hero of *From the Earth to the Moon*, who told the crowds in Tampa Town that the journey to the moon 'must be undertaken sooner or later; and, as for the mode of locomotion adopted, it simply follows the great law of progress'. Because mankind had advanced from moving on all fours to travelling in trains, it was certain that one day men would take off into space. 'Yes, gentlemen,' says the orator, 'I believe that the day is coming when the great ethereal ocean enveloping the universe can be crossed, when we can take passage for the Moon, for the planets, for the stars, as we now take passage from New York to Liverpool, as easily, as rapidly, as safely!'

Another powerful factor in the Verne stereotype was the super-abundance of factual details that obscured the almost total elimination of normal human experience. Verne usually saw to it that the complications of politics or passion would never hamper the swift progression of his plots; and in this way a sensible formula for juvenile fiction took on utopian elements of perfection and power that appealed to the adult reader. Emphatic language, confident projections and detailed explanations were the means by which Verne combined a necessary verisimilitude with the benediction of scientific certainty. He began by presenting technological possibilities through the adventures of heroic individuals. For their sphere of action he chose the whole world, or even the depths of space, and he ensured that his heroes were scientists or engineers who would have the double function of representing nineteenth-century technology and of convincing the reader of the possibility of a journey to the centre of the Earth or of the practicability of a vast submarine. 'There is a powerful agent,' says Captain Nemo, 'which conforms to every use, and reigns supreme on board my vessel. Everything is done by means of it. It lights it, it warms it, and it is the soul of my mechanical apparatus. This agent is electricity.'

Verne had an exhaustive explanation for everything, even for his most eccentric invention, the perambulating steam-elephant in *The Steam House*: 'Between the four wheels are all the machinery of the cylinders, pistons, feed-pump etc., covered by the body of the boiler. This tubular boiler is in the fore part of the elephant's body, and the tender, carrying fuel and water, in the hinder part.' In this way, by grounding his stories in scientific facts, Verne was able to obtain the desired illusion of reality. His characters talk like members of the Royal Society and they pursue their objectives in what appears to be the real world; but it is clear that, once Professor Arronax descends into the *Nautilus*, the reader enters the realm of fantasy and the process of wish-fulfilment begins. So, in his artful way, Verne used the circumstantial evidence he scattered through every story in order to conceal an elaborate mimicry of contemporary experience. He saved his stories from the dangerous limitations of time and reality by placing his heroes in the expanding universe of nineteenth-century expectations; and since he was always careful to maintain the closest connexion between theory and practice, as well as between contemporary ambitions and their projected realizations, everything appeared self-evident and necessary in the deceptively simple conditions of his stories.

The marvels of science, however, depend upon the magic of myth and fairytale – a fact that explains the continued popularity of Verne with the film-makers. The Icarus adventures of Robur in *The Clipper of the Clouds*, like the Poseidon role of Captain Nemo, are potent manifestations of the eternal desire that the power of man over nature shall always be as instant and as absolute as his will. Thus, Verne's principal heroes succeed in everything they do, because they have had the extraordinary good luck to obtain or to inherit the immense fortunes required to develop their projects. Captain Nemo draws on the treasury of his family for the building of the *Nautilus*; Phileas Fogg literally throws money away; Dr Sarrasin and Professor Schultze divide the begum's fortune of £1,000 million; Matthias Sandorff inherits £50 million from a wealthy patient; and the whole world contributes to the cost of building Barbicane's *Columbiad* – except the mean-souled British, of course – because 'the sum required was far too great for any individual, or even for any single state to provide the requisite millions.'

The more memorable characters – Nemo, Robur, Phileas Fogg, Barbicane, for example – are solitaries. They pursue their objectives with unrelenting obsession and with unfailing success;

and the accounts of their exploits are set aside from the rest of human experience—beneath the waves in a prodigious submarine, high above the clouds in a flying machine, hurled towards the Moon in a space projectile. As each story unfolds, the scale of the enterprise widens to take in the whole world and the imagination rises to grandiose images of human destiny: Nemo unfurls his black flag and takes possession of the South Pole in his own name; the astronauts of the *Columbiad* gaze in silence at the barren surface of the Moon; and Robur stands beneath the seventy-four whirling screws of the *Albatross*, proclaiming that he is 'master of the seventh part of the world, larger than Africa, Oceania, Asia, America, and Europe, this aerial Icarian sea, which millions of Icarians will people one day'.

These climactic moments are the sacramental celebrations peculiar to the religion of progress. In a dramatic and ritual manner they invite the reader to rejoice in the world-changing capabilities of the new technologies; and this impression of a potent union between science and nineteenth-century society finds supporting evidence in the repetitive language and the detailed descriptions that never fail to appear whenever Verne starts on one of his lengthy accounts of some great engineering construction. There is a liturgical and invocatory quality in many of these episodes, for Verne made a practice of dwelling on the essential factors of weights, ratios, tensile strengths, speeds, measurements of every kind and form. The chapter on the casting operations in *From the Earth to the Moon* notes each stage of the process—68,000 tons of coal, 1,200 furnaces, 60,000 tons of iron. Then the gun-shot gives the signal to vent the molten metal, and through the roaring of the furnaces Verne tells the reader that 'it was man alone who had produced these reddish vapours, these gigantic flames worthy of a volcano itself, these tremendous vibrations resembling the shock of an earthquake, these reverberations rivalling the hurricane and the storm; and it was his hand which precipitated into an abyss, dug by himself, a whole Niagara of molten metal.' The Olympian observer has used the language of the gods to comment on the powers of mankind in a technological epoch; and on another occasion, when Robur comes down from the clouds like Zeus out of high heaven, Verne legislates for the world: 'Citizens of the United States, my experiment is finished, but my advice to those present is to be premature in nothing, not even in progress. It is evolution and not revolution that we should seek. In a word, we must not be before our time.'

In this way Verne worked out story by story what he felt to be the appropriate mythology for his times. In the three parts of *The Mysterious Island*, for instance, he went through the cycle of human existence from the beginning of civilization to the end of the world. When his balloonists descend on their desert island, the primordial moment begins with the announcement that 'they had nothing save the clothes which they were wearing ... not a weapon, not a tool, not even a pocket-knife'. And then Prometheus, in the person of the engineer Cyrus Harding, creates a new cosmos out of his knowledge of science and the materials available on the island: 'The engineer was to them a microcosm, a compound of every science, a possessor of all human knowledge. It was better to be with Cyrus in a desert island than without him in the most flourishing town in the United States.' Throughout the rest of the story Verne applies science with a conspicuous didacticism to the Robinson Crusoe situation of the castaways. More fortunate than their predecessor they begin by making fire, then they manufacture bricks, domesticate animals, produce iron implements and high explosives, and end by making electricity. Science had transformed the Crusoe condition in the same impartial way as scientific knowledge imposed the conclusion that ' ... some day our globe will end, or rather that animal and vegetable life will no longer be possible, because of the intense cold to which it will be subjected.' Thus, the journey through the world of the Verne stories becomes a descent into the subconscious to the source of symbols and desires; for Verne renews the vigour of the ancient myths by projecting his characters into a self-perpetuating world of eternal progress and increasing power. As Hetzel said in the early years of their collaboration, Jules Verne was going to rewrite the story of the universe.

As he grew older, Verne became increasingly unsure about the future of the universe; and the anxieties he began to express about the lethal potentiality of the sciences at the end of the

1870s followed the descending curve of his once unqualified optimism. Verne had seen the arrival of the iron-clad warships and the development of new means of warfare, and like many of his contemporaries he came to realize that the applied sciences were a mixed blessing. He first presented this basic dilemma, as he saw it, in the unusually elaborate action in *The Begum's Fortune* of 1879 which he divided between the virtuous activities of Dr Sarrasin, who founds a hygienic utopia at Franceville in the United States, and the machinations of the wicked Professor Schultze of Jena who establishes his despotic society at Stahlstadt thirty miles away from his rival. Schultze was the first of the demon scientists, and his secret weapon for use against Franceville was a monster gun of 300 tons: 'The general opinion was that Professor Schultze was working to complete a terrible engine of war of unprecedented power, which was intended to assure universal dominion to Germany.' It was designed to fire carbonic acid gas shells; and when these explode on impact, '... an enormous volume of carbonic acid gas rushes into the air, and a cold of a hundred degrees below zero strikes the surrounding atmosphere. Every living thing within a radius of thirty yards from the centre of the explosion is at once frozen and suffocated.'

This was in the style of the imaginary wars of the future that had become familiar matter in France, Germany and the United Kingdom after the *Battle of Dorking* had in 1871 given the first international warning of the change in the conditions of warfare. During the last three decades of the century, as the demand for tales of the future went on growing, more and more writers learnt from Verne how to exploit the stock-in-trade of the ambitions of the age. A new race of journalists and popular writers began to provide complacent stories about the wonders-to-come for the mass of literate citizens in Europe and the United States. They traded in the terrors and delights of technological society with their exciting stories of space travel, future worlds, death rays, demon scientists, imaginary wars, flying machines and the rest of the marvels of science fiction. Their base lay in the popular newspapers and illustrated magazines that began to appear towards the end of the 1880s, since it suited the interests of the publishers and the writers to start by placing a story as a serial in a periodical and then to bring it out as a book.

The French were the first to exploit this system in the very lucrative and life-long association between Jules Verne and Hetzel; and the French were again the first to find profit in the new popular interests when the able writer and publicist Louis Figuier founded *La Science illustrée* in 1887 and thereafter delighted his readers week by week with the stories of Louis Boussenard, Camille Flammarion, Camille Debans, Albert Robida, and other pioneers of science fiction in France.

Most of the writers who followed in the tradition of Jules Verne were content with variations on the favourite stereotypes of the day. Thus, André Laurie produced his version of the interplanetary adventure theme in *Les Exilés de la terre* in 1887 and in the following year he turned to the romance of engineering in *De New-York à Brest en sept heures*, in which he described the construction of the first pipe-line under the ocean to convey oil from the United States to Europe. A far more interesting innovation appeared in *L'Eve future* (1886), in which Villiers de L'Isle-Adam had the original idea of using Thomas Alva Edison to create a gynodroid (one cannot call it a female android nowadays) who is programmed to give the English aristocrat, Lord Ewald, the love and companionship the poor wretch cannot find with real women. The preface revealed the intimate connexion that then existed between the popular image of the scientist and the miracles of science fiction. The author noted the many inventions that had given Edison a world-wide reputation: 'In America and in Europe a legend has thus grown up in the popular imagination about this great citizen of the United States. He goes by such fantastic titles as "the Magician of the Century, the Wizard of Menlo Park, the Father of the Phonograph". The most understandable enthusiasm in his own country and abroad has given him a kind of mysterious quality rather like that of many spirits.'

So, writers everywhere drew upon the deposit of contemporary credibility for their accounts of the future; and between the extremes of peace and war they roamed at will, dealing in the exciting rhetoric and heroic deeds of 'the next great war' or giving their readers the utopian delight of living in the most desirable world of the twentieth century. The most striking illustration of this general practice appeared in the extraordinary fantasies of Albert Robida, the most gifted and original artist in the history of science fiction. His singular talent, remarkable even in that age of prophets, lay in his ability to imagine a future period in which all the anticipations of his day would decide the way of life in a self-explanatory world of the twentieth century. In 1882 he began the serial publication of *Le Vingtième Siècle*, in which his exceptional imagination created a complete society at work and play in the year of 1952: angry wives upbraid their husbands by *le téléphonoscope*; women wear short dresses designed for use in flying machines; families watch the latest news from Africa on the television; there are food factories, submarine towns, aerocabs, and underwater sports. All the elements in the Robida projection coincided in the natural unity of a feasible society; and that same coherence of the imagination decided the application, but failed entirely to foresee the consequences, of the military technologies he was the first to describe in his spectacular fantasies on twentieth-century warfare.

The first version of *La Guerre au vingtième siècle* appeared in *La Caricature* on October 27th, 1883, and in 1887 Robida brought out a revised edition in an illustrated album that is now a prized item with collectors. The text begins with the outbreak of the Great War of 1945; and in a mood of ironic comedy Robida explains that hostilities have followed on a commercial dispute, because 'the civilized nations nowadays dictate their commercial treaties to the sound of gunfire'. The familiar devices of modern warfare are already at work in Robida's logical mind; flying machines bomb enemy cities and attack enemy tanks that run for cover into woods; the troops of the chemical engineering battalions use poison gas and fire off shells containing bacterial preparations of deadly illnesses; flame-throwers operate against defensive positions; submarines and underwater troops seek out the enemy warships. All these lethal activities take place in a no-man's land of the imagination. Although Robida describes the obliteration of cities, his sense of innocent amusement holds him back from imagining the consequences of total warfare in his future Europe. With great good humour he relates how his hero sets fire to a chemical warfare depot, with the amusing result that '... the enemy hospitals had to look after 179,549 civil and military casualties, and that the mixing of all these poisonous concoctions produced a remarkable and absolutely new illness. This was developed by doctors all over Europe, and is now known as Molinas Fever after the name of the inventor. The spot where it all began has remained a most unhealthy place.'

This grotesque failure in anticipation was the norm for all but a handful of the imaginary wars that appeared before the bitter lessons of Verdun and the Somme taught the Europeans the consequences of their cherished technologies. Because the great innovations of the nineteenth century had apparently gone forward in a logical fashion – bigger, better, faster – and because the dominant belief in a benign evolutionary progression had underpinned the expectation of continuous social improvement, it was the most natural thing in that world for writers to think that the new technologies would always have beneficent results for the fortunate citizens of the great urban societies. The projections of Robida and the other authors of imaginary wars depended on a dangerous difference between imagination and intellect that worked against the dispassionate analysis of future possibilities. The bias of their minds allowed them to entertain the contradictions of all-change and no-change in their visions of the great wars and of the immensely lethal armaments that could never have any disastrous effect upon the complex structure of an industrialized world.

As the tale of the future grew in popularity during the last two decades of the nineteenth century, the imaginary wars and the new science fiction stories became the opium of the urban masses. In the United Kingdom the astute publishers of the first popular newspapers and magazines – Newnes, Pearson, Harmsworth – repeated the practices of their French contemporaries. They made a feature of short stories and serials about future wars and glorious victories, and they found space for science fiction variations on the staple themes of space adventure and all-powerful scientists. Who now remembers the many publications of George Griffith, Robert Cromie, Fred T. Jane, William LeQueux, and Louis Tracy?[3] They were all, as they said in those days, gentlemen of the press; and they saw to it that their yarns gave their readers the standard mixture of violent action, marvellous inventions, and frequent utopian conclusions. So, Robert Cromie, who ended his days as editor of a Belfast newspaper, turned out a series of future war stories, and in *The Crack of Doom* of 1895 produced one of the more remarkable of the demon scientists who make war upon the human race. His villain has discovered that '... one grain of matter contains enough energy, if etherized, to raise one hundred thousand tons nearly two miles'. Before he starts on his plan to destroy the world, the wicked scientist demonstrates his powers by annihilating a French fishing fleet:

> 'You will hear the explosion in ten seconds' ... Then the seas behind us burst into a flame, followed by the sound of an explosion so frightful that we were almost stunned by it. A huge mass of water, torn into a solid block, was hurled into the air and there it broke into a hundred roaring cataracts. These fell into the raging cauldron that seethed below. The French fishing fleet had completely disappeared.[4]

Another series of imaginary wars came from Louis Tracy, one of the new wave of journalists, who had acted as intermediary when Newnes sold the *Evening Times* to the Harmsworth brothers. In his story of *The Final War* of 1896 Tracy tapped the contemporary interest in Andrew Carnegie's suggestion in 1893 that the United Kingdom and the United States should enter into a political union. All this comes to pass when the Americans and the British defeat the Franco-German forces; and after final victory has given the Anglo-Saxons the mastery of the world, the author proclaims the evolutionary doctrine that,

> as life becomes more complex and harder grows the struggle, there is no escape for peoples not fitted to bear its strain. The Saxon race will absorb all and embrace all, reanimating old civilizations and giving new vigour to exhausted nations. England and America – their destiny is to order and rule the world, to give it peace and freedom, to bestow upon it prosperity and happiness, to fulfil the responsibilities of an all-devouring people; wisely to discern and generously to bestow.[5]

The flattering prophecy of a world that would be for ever Anglo-Saxon may have been a patriot's dream. It was certainly the kind of happy ending given to many of these imaginary wars, and it was undoubtedly a shrewd application of the Harmsworth formula for giving the readers what they liked.

This client relationship between press and public became a major factor in the popular journalism of the 1890s. Across the Atlantic, for example, American authors gave their readers what they liked – imaginary wars fought with European ferocity, but in different campaigns that revealed the various forces at work in the United States. In the *Last Days of the Republic* by P. W. Dooner the armies of Imperial China overrun the sweet land of liberty from sea to shining sea. In the anonymous *Fall of the Great Republic* and in *The Stricken Nation* by H. G. Donnelly the ships of the Royal Navy bombard all the ports along the eastern seaboard, and the Americans have to submit to the dreadful indignities of an imaginary British occupation. But the tide turns in *The Dynamite Ship* by Donald McKay, when Irish revolutionaries attack London with dynamite guns; and by the end of the 1890s the two nations had become allies in the then fashionable libretto, popular on both sides of the Atlantic, that looked forward to a successful Anglo-American war against the rest of the world. In 1898 Stanley Waterloo described in *Armageddon* how an inventor used his bombing plane to assist the fleets of the Anglo-American-Japanese alliance against the French and Russians; and in another forecast of 1898, *Anglo-Saxons Onwards!* by Benjamin Davenport, the author concludes with the slogan of 'Anglo-Saxons, Onwards! America and England for ever! Onward for our race and nations!'

Whenever some matter caught the public interest – canals on Mars or the Women's Rights Movement, for instance – then there would follow stories about interplanetary travel or about the role of women in a future society. In the 1880s, as the campaign to remove the injustices suffered by women had begun to achieve some success, the tale of the future gave the reformers and the die-hard conservatives an opportunity to write in support of their convictions. The most explosive diatribes came from Percy Greg, a political journalist and a man of violent extremes. In his space story, *Across the Zodiac*, Greg brought the maladies of his time to their logical conclusions in the highly advanced society on Mars; and he showed that the Martians had learnt one major lesson from their history – women are not equal to men:

> The Equalists were driven from one untenable position to another, and forced at last to demand a reduction of the masculine standard of education to the level of feminine capacities. Upon this ground they took their last stand, and were hopelessly beaten. The reaction was so complete that for the last two hundred and forty generations the standard of female education has been lowered to that which by general confession ordinary female brains can stand without violent injury to the physique.

Another Victorian male, William Delisle Hay, solved the problem of women's role in the technological Eden of the future by giving them the appearance but not the reality of power; and the more he transformed the contemporary world in his vision of the planetary society of *Three Hundred Years Hence*, the more firmly he kept women in their place in his grand Victorian millennium. His world state of the twenty-second century is based on '... that Unity which was the very essence of Progress, the very gospel of the Future'; and the supreme manifestation of that unity is the election once every ten years of the Empress of the Earth, 'the Ideal of Beauty, the type of Innocence, the emblem of United Humanity, and the incarnation of Poetry, Sentiment, Romance, and Love.' What would Mrs Gandhi and, indeed, Mrs Margaret Thatcher say about this statement from a conservative Victorian male?

> Thus, in the system of government set up by united Humanity, Woman finds her station. Nominally the head of all government in the abstract, yet the Empress possesses not one particle of actual power over the decisions of Oecumenic Wisdom. This is as it should be, for the faculties of women are not capable of following out the highest intellectual processes, since she has a less degree of pure Reason than the man. Yet is Woman most emphatically fitted to display in herself the ornamental part of government, while, her sway

over man being gained by appeal to his passions rather than his intellect, she is manifested as the proper director of all that relates to domestic life. It is undoubted that Woman's virtues shine forth most clearly in the refining influences she exercises over the home economy and the domestic circle, and as the unexpressed and indefinite laws of society in past ages were the outcome of Woman's wit, so we have given her the power of forming those laws into an actual code, based on pure and noble principles and of carrying them into practice with sure and certain effect.[6]

The debate about the place of women in society ran its course in the usual way of futuristic fiction, and writers contrived to show that justice for women would create the best or the worst of worlds: Sir Walter Besant was all against equality in *The Revolt of Man*, 1882; E. B. Corbett in *New Amazonia* and H. R. Dalton in *Lesbia Newman*, both published in 1889, were all for the emancipation of women; and there were other versions of the argument in *Gloriana; or the Revolution of 1900* (1890) by Lady Dixie, *The Strike of a Sex* (1891) by G. N. Miller, *Mercia the Astronomer Royal* (1895) by A. G. Mears, *The Rev. Annabel Lee* by R. Buchanan (1898). For the reputation of the men the last words can be left to Sir Julius Vogel who wrote in a rational and restrained manner in his prophecy of 1889, *Anno Domini 2000; or Woman's Destiny*. He reported that the greatest advances of the twentieth century would be

> ... the astonishing improvement of the condition of mankind and the no less striking advancement of the intellectual power of woman. The barriers which man in his own interest set to the occupation of women having once been broken down, the progress of woman in all pursuits requiring judgement and intellect has been continuous; and the sum of that progress is enormous. It has, in fact, come to be accepted that the bodily power is greater in man, and the mental power larger in woman. So to speak, woman has become the guiding, man the executive, force of the world.[7]

The common theme in these stories shows how matters of the moment provided the material for a profusion of futuristic narratives during the last two decades of the century. At the same time a change took place in the presentation of these stories, as more and more writers sought to place their themes in a moral or social setting. So, a growing search for some universal coherence in the scheme of things or for some final solution to all social problems imposed the common pattern of an international, sometimes even an interplanetary, order on most of the utopian prophecies and many of the science fiction stories and imaginary wars that appeared after 1880. In the imperialistic manner the dogmas of the political propagandists and the necessities of the nation state led to visions of a stable and united world dominated by one or other of the great powers; and even in the more extravagant science fiction stories the action took place in the context of a watching and waiting world, rather in the style of the modern television serials.

In 1888, for example, André Bleunard produced one of the more charming fantasies of the century, *La Babylone électrique*, which was 'the history of an expedition to restore Ancient Babylon by the power of electricity', a project that was supposed to attract support throughout the world. In 1889 the prolific American writer Frank Stockton related in *The Great War Syndicate* how Yankee ingenuity and business enterprise would provide the secret weapons needed in the future war to make the United States the leader of all nations on earth. And in 1897 two writers chose to reverse the hallowed sequence of the space travel stories by bringing their Martians to our planet. Whilst Wells's most original account of the first great interplanetary conflict in the *War of the Worlds* was appearing in *Pearson's Magazine*, Kurd Lasswitz published his version of a peaceful Martian visitation in *Auf zwei Planeten*. Both men had started from the same Darwinian proposition – that any major change for the worse in the circumstances of life on a planet of the solar system could force the members of any highly advanced technological civilization to seek out more favourable conditions for their survival in our world.

This striking application of European colonial experience was the logical development of the concept of one world that Sebastien Mercier had first introduced in the late eighteenth century; and the evident desire to present the various themes of futuristic fiction in a more comprehensive manner and in the context of a world society was the immediate consequence of theory and experience. Evolutionary ideas, histories of progress, many theories of urban society – political and sociological – had all confirmed and enlarged the experience of life and work in continents that the new means of communication were joining together in ever more complex and intimate relationships.

Evolutionary theories, the idea of progress and the capabilities of a technological society were central to the thinking and to the success of H. G. Wells. In the space of 10 years, from the publication of *The Time Machine* in 1895 to the ambitious programme for *A Modern Utopia* in 1905, Wells caught the interest of his contemporaries with a series of most original stories about time, space, future wars and future societies. The foundations of his extraordinary career as the first prophet laureate to the world lay in the special circumstances of the 1890s, since Wells applied his very original talents to the favourite themes of futuristic fiction in that decade. One example of Wells's success with his readers appears in the publishing history of the *War of the Worlds*; for the unscrupulous treatment he suffered from the editors of the New York *Journal* and the Boston *Post* shows only too clearly now Wells had caught the general interest with a most imaginative variation on the universally popular stereotype of the imaginary war of the future.

The episode began in the September of 1897, when Arthur Brisbane became editor of the *Journal* on terms that paid him by results. Within twelve months he had increased sales tenfold and had made the *Journal* notorious. His first sensational enterprise was to buy the American serial rights for the *War of the Worlds*, which had been published simultaneously in *Pearson's Magazine* and in the American *Cosmopolitan* between April and December in 1897; and he arranged for the Boston *Post* to carry the version of the story that would first appear in the *Journal*. Brisbane had not told Wells that he intended to make large-scale changes in the text so that the New Yorkers could have the excitement of reading about a devastating invasion of the United States. So, he began on December 15th, 1897, by shifting the location of the Martian invaders from southern England to the New York area – in the *Post* version the scene changed to

Boston – and editorial staff introduced specially written episodes that enlarged on the destructiveness of the Martian fighting machines – a puff of smoke and the Brooklyn Bridge is destroyed, another puff and Columbia University is in ruins. In the *Post* adaptation the Martians advance from Concord to Lexington, and their heat rays destroy the hallowed monuments of American history. Of course, H. C. Wells, as the two editors called him, protested against this infringement of copyright in the land of the pilgrims' pride, but he had no hope of redress. All he could do was to write to the American *Critic*, explaining what had happened and 'disavowing any share in this novel development of the local color business'. There was more to come. The *Post* version of the Wells story had been so successful that Arthur Brisbane at once commissioned another writer, Garrett P. Serviss, to prepare a sequel. This was *Edison's Conquest of Mars*, in which, the *Journal* claimed, without giving any evidence, Thomas Alva Edison had had a hand; and the first number appeared immediately after the end of the Wells serial on January 12th, 1898.

The preposterous story opened with Lord Kelvin, Roentgen, Edison and other scientists working to perfect the machines that would enable the men of Earth to carry the war to Mars:

> Suddenly cablegrams flashed to the government at Washington, announcing that Queen Victoria, the Emperor William, the Czar Nicholas, Alphonso of Spain, with his mother, Maria Christina; the old Emperor Francis Joseph and the empress Elizabeth ... and the heads of all the Central and South American republics were coming to Washington to take part in the deliberations, which, it was felt, were to settle the fate of earth and Mars.

The banal episodes in *Edison's Conquest of Mars* serve to show that Wells attracted the attention of the world because he was the best and the most inventive of many writers who competed for space in the periodicals and the illustrated magazines. He stood alone, pre-eminent in a crowded field. As G. K. Chesterton noted, Wells's 'first importance was that he wrote great adventure stories in the new world the men of science had discovered'. Another of his contemporaries, Arnold Bennett, considered that in the scientific romances Wells reached the level of epic poetry and he was certain that 'no future novelist will be able to "fudge" science now that Wells has shown how it can be done without fudging'. Wells became the favourite seer of the age, because he combined a professional knowledge of science with an unusually original and fertile imagination. He used his gifts to explore all the areas of the future that most interested his contemporaries – Darwinian ideas about human evolution in *The Time Machine*, remarkable variations on Darwinism and the theme of technological warfare in the *War of the Worlds*, the despotism of a future society in *When the Sleeper Wakes*, a most famous tale of space travel in *The First Men in the Moon*; and in *Anticipations* he produced the first popular prediction about the course of social and scientific developments in the twentieth century.

It was an unprecedented achievement. In seven years Wells had made himself the true begetter of modern science fiction and a founding father of modern practice in social and technological forecasting. In 1902 he read a paper before a crowded meeting of the Royal Institution. The subject was 'The Discovery of the Future', and he told his audience that 'an inductive knowledge of a great number of things in the future is becoming a human possibility. I believe that the time is drawing near when it will be possible to suggest a systematic exploration of the future.'[8] By 1905, the year Verne died, Wells had given the world his scheme for a planetary society in *A Modern Utopia*, and as the high priest of the future he prophesied for his people: 'Here we are, with our knobby little heads, our eyes and hands and feet and stout hearts, and if not us or ours, still the endless multitudes about us in our loins are to come at last to the World State and a greater fellowship and the universal tongue.'[9]

PART TWO

The Variant Forms

5

Ideal States and Industrial Harmonies

The nineteenth-century tale of the future reached the peak of achievement in the varied and inventive stories of H. G. Wells. His special contribution to the development of the genre was to find original applications for those themes of future warfare, future despotism, future utopias and future technological advances that had held the interest of readers from Cousin de Grainville's history of *The Last Man* in 1805 to the Wellsian vision of an empty world at the end of *The Time Machine* in 1895.

Looking backward from the achievements of Wells about the turn of the century, it becomes evident that a succession of new themes and new styles has marked the evolution of futuristic fiction – from the ideal victories and the ideal achievements of the Patriot King in *The Reign of George VI* in 1763 to the mass production of scientific romances that began with the new popular press in the 1890s. The constant and close relationship between the interests of contemporary society and the projections of futuristic fiction encouraged the separate development of the science fiction story, the tale of 'the next great war', and the ideal states of the future. This differentiation of specific themes in the genre depended on a corresponding division in authorship. The story of the war-to-come, for example, first began to influence the thinking of nations after the worldwide notoriety of the *Battle of Dorking* in 1871. That story, written by a colonel of the Royal Engineers, established a special form of futuristic fiction that became a favoured means of communication with senior officers, cabinet ministers, and military correspondents in France, Germany, and the United Kingdom. The very different development of science fiction, however, came about through the commercial enterprise of publishers and the specialized writing of professional authors. Verne, Wells and the journalists of the 1890s made it their business to understand the requirements of the market, and they worked with the publishers to give their readers what interested them.

The ideal state of the future and its black opposite, the dystopia, had a parallel but very different history of development. These composite answers to the state-of-the-nation question have always had a special attraction for writers with decided views on matters of great social concern – poets like Lamartine and William Morris, gifted and original writers like Anthony Trollope and W. H. Hudson, social theorists like Edward Bellamy and Theodore Hertzka. To describe what they wrote and how they wrote is to present the history of a unique and most influential form of futuristic fiction in the course of an evolutionary development that goes from the ambitions of Sebastien Mercier in *L'An 2440* to the detailed proposals of H. G. Wells in *A Modern Utopia* in 1905.

The initial phase in this eminently social literature fell within the 100 years between *L'An 2440* of 1771 and the sudden flood of utopias that began about 1871. Thereafter, from 1871 to 1914, the main period saw the growth of the largest, most varied and most influential body of utopian fiction in the history of the genre. The characteristic form was – with some notable exceptions – the ideal state of the future in which anticipation was the prophetic image of the aspirations of industrial society. The progressive and rational utopia was the self-confident manifestation of the Promethean period in an era of technological progress; and the doctrinal base for these visions of the happiness-to-come was the new belief in the immense powers of a technological society – eternal, without measure, self-perpetuating, dominating and directing the forces of nature to the advantage of mankind. The architects of the constructive ideal state had heard destiny knocking on the door to the future, and according to the vigour of their imaginations or the pressure of their political interests they wrote their expectations into visions of a better world, descriptions of improved metropolitan conditions, or practical designs for a new social order. Their prophecies record the ceaseless dialogue of urban society with its best self which has continued without pause, but in other forms, throughout the twentieth century; for all these utopian prognostications synchronize with the social, political and technological advances of the last two hundred years, for ever moving towards the Omega point in time-to-come when the nation or the whole world is supposed to reach the predestined goal of an ideal existence.

These ideal states of the future are, in fact, points on the graph of progress. They mark the swift ascent from the rudimentary calculations of the last quarter of the eighteenth century to the statistics and the more expert predictions that were central to the first essays in technological and social forecasting at the end of the nineteenth century. The scale of this development is apparent in the difference between the theorizing of Malthus in the *Essay on the Principle of Population* and Wells's readiness to predict the diffusion of great cities, many advances in warfare and locomotion, changes in education and religion in *Anticipations* in 1901; and it is equally evident in the advance from the visions of steam vehicles and flying machines in the early decades of the nineteenth century to the original and often perceptive account of the coming of atomic energy in Wells's *The World Set Free* in 1914.

As the industrial nations moved ever further away from the immemorial traditions and practices of the old agrarian economies, a comparable change transformed the conventions of utopian fiction. The discovery of Australia and the many expeditions to the unknown interior of the great continents had filled in the empty spaces on the map of *Terra Nondum Cognita*, the venerable home of ideal communities ever since the time of More and Bacon; and for these reasons the majority of propagandists abandoned the device of the terrestrial utopia in favour of a mass migration to new locations in time-to-come. The few who continued with the old tradition found it more and more difficult to generate the geographical realism that was required to support the factual account of life and work in the ideal state.

There were various solutions, and one of the more ingenious appeared in *Armata*, published in 1817 by the greatest of Scottish advocates and one-time Chancellor of the Exchequer, Lord Erskine. The story begins with the consecrated account of the sea voyage to distant lands, the hero sailing from New York in the good ship *Columbia*, bound for China by way of New South Wales; and in the customary manner a storm sweeps the vessel off the map, driving it along a narrow channel that, like an umbilical cord, joins Earth to the unknown satellite planet of Armata. The opening passages compose an epitaph for the old-style utopia, since the narrator explains the previous failures to discover Armata by the fact that the planet '... had a ring like Saturn, which, by reason of our atmosphere, could not be seen at such an immense distance, and which was accessible only by a channel so narrow and so guarded by surrounding rocks and whirlpools, that even the vagrancy of modern navigators had never fallen in with it.' The careful phrases expose the problem of trying to authenticate the terrestrial utopia in a way no longer countenanced by geography. One solution was to give up the unequal struggle for a measure of realism in favour of unashamed make-believe. This had been the solution of James Lawrence who began *The Empire of the Nairs* in 1811 by stating that the citizens of his perfect society were to be found on the Malabar coast of Indostan: 'The mighty empire, which is ceded to them in this novel, like Brobdingnag and Lilliput, will be found in no book of geography ... The Paradise of the Mothersons is merely ideal.'

Another and more realistic solution, which later became a popular method of creating a geographical setting for the imaginary land, was to discover a Lost Race in some remote part of the world. In a French story of 1810, *Le Vallon aérien*, the narrative opened with the adventures of an aeronaut who explores the Pyrenees by balloon and happens on an ideal community in a hidden valley sealed off from all contact with the outside world. Again, in a similar English utopia of 1820, *New Britain*, the sub-title promised a full account of the new land '... discovered in the vast Plain of the Missouri, in North America, and inhabited by a People of British Origin, who live under an equitable System of Society, productive of peculiar Independence and Happiness.'

Yet another answer to the problems of utopian fiction was a revised version of the traditional journey to the Moon; and the most original of the new interplanetary voyages came from George Tucker, the first professor of Moral Philosophy in the University of Virginia, who published his *Voyage to the Moon* in 1827, the year Edgar Allan Poe was a student there. Later on Poe borrowed the documentary and factual style of Tucker's narrative for his lunar story *Hans Pfaall*; but Poe chose to make the space journey by balloon, whereas Tucker developed a variant on the anti-gravitational device of Swift's flying island of Laputa, and in this way the professor became the first of many writers who have since taken off into space thanks to the extraordinary properties of some mysterious substance. In Tucker's story a mystic Brahmin from Benares had discovered a rare metal, *lunarium*: 'This metal, when separated and purified, has as great a tendency to fly off from the earth, as a piece of gold or lead has to approach it.' The Brahmin uses the metal to drive his spacecraft, 'a copper vessel, that would have been an exact cube of six feet, if the corners and edges had not been rounded off.' At the turn of a screw the space traveller activates the *lunarium*, and the vehicle lifts off to begin the forty-eight-hour journey to the Moon.

Although these writers had laboured to preserve the traditional utopia from the encroachments of the discoverers, their stories were not in any sense a turning towards the past. On the contrary, like the contemporary authors of futuristic utopias, they were concerned to pick out the factors that would improve the state of society and the condition of life. In the ideal Britain of Erskine's *Armata* the projected advances in agriculture depend upon the discoveries of science; for 'the science of agriculture is by no means at its heights, and in the almost miraculous advance of chemistry new means may be found, from the concentration of known composts and the discovery of new, to lessen the cost of culture and to increase its returns.'[1] And in a similar way in *New Britain* the ideal social system depends upon technology: 'The progress of the New Britons in mechanical science is very great, and its application has been in proportion.

They apply all the generally useful machinery of Europe, in addition to their own inventions. Machinery does not here, by the rapidity with which it completes the objects of labour, enrich a few and deprive many of the means of procuring the necessaries of life, as are its miserable effects in other countries.'[2] This fusion of archaic form and prophetic content appears most strikingly in the *Voyage en Icarie*, a famous and influential utopia of 1840, in which Étienne Cabet set out the attractions of an organized, technological and communist form of society. Cabet located his ideal state somewhere in the Indian Ocean and, with a fine disregard for probabilities, made it accessible to all visitors. Icarie is France – but with a different and a happier history. A revolution has swept away all the old inequalities and injustices; and the new society emerges, dedicated to the principle that 'the modern age is not only the epoch of Democracy and Equality but also that of *industry* and *production*'. The italics emphasize Cabet's conviction that the rational pursuit of happiness must lead to a communist state in which the perfect society will enjoy all the benefits of the most advanced sciences. Thus, the Icarians have mechanized the construction industries; they have discovered a source of power far more efficient than steam; they have mastered the problems of dirigible flight and their practice of using machines wherever possible has so changed life on the land that '... the task of the farmer is almost reduced to that of an intelligent director and an enlightened manager'.

What Erskine and Cabet had sought in their revised versions of the terrestrial utopia was the opportunity to exploit an element of allegory and parallelism in order to concentrate attention on their social theories. By transposing their arguments from the immediate and contemporary to imaginary and ideal societies they hoped to demonstrate the operation of those universal laws that must decide the progress of mankind. Unfortunately this left their propaganda awkwardly placed between most obvious allusions to the history of their own countries and the contrived illusions of their undiscovered lands. Distance could no longer lend enchantment to the view of these faraway nowheres. In the evolutionary conditions of the nineteenth century the imaginative presentation of social, political and progressive ideas required a prospective view of society that the tale of the future alone could project in any truly satisfactory manner. And so the terrestrial utopia went into the rapid decline from which it has never recovered, since the constraints of the old form had introduced an inescapable contradiction into all accounts of the imaginary land with the significant exception of the new-model Erewhonian satires that made a virtue of literary convention. The satirists found the appropriate context for the moral concerns and the limited objectives of their dystopias by vanishing below the surface of the earth into the subterranean world of Bulwer-Lytton's *The Coming Race* or by going right off the map into the nowhere of Samuel Butler's *Erewhon*. Even there the onward movement of the age had its effect. The timeless tale of literary tradition became a future-relative history in which the author foresaw the lamentable consequences of nineteenth-century failings. So, Bulwer-Lytton showed that the subterranean Vril-ya had realized all the dreams of his contemporaries and that their way of life was

> ... immeasurably more felicitous than that of the super-terrestrial race, and, realizing the dreams of our most sanguine philanthropists, almost approaches to a poet's conception of some angelical order. And yet, if you would take a thousand of the best and most philosophical of human beings you could find in London, Paris, Berlin, New York, or even Boston, and place them as citizens in this beatified community, my belief is that in less than a year they would either die of *ennui*, or attempt some revolution by which they would militate against the good of the community, and be burnt into cinders at the request of the Tur.[3]

These satires were no more than occasional and dissenting interjections in the century-long debate about the beneficial tendencies of the age. Throughout the hundred years before the outbreak of the First World War the dominant utopia of the future presented a regular succession of progressive ideas, and these decided the general attitudes and the major propositions to be found in a unique conjectural literature that was both European and North American in origin. Four quite exceptional conditions separated these ideal states of the future from the practices of the old terrestrial utopias. First, most of them carried on and extended the innovations of *L'An 2440* and *The Reign of George VI* by placing the nation at the centre of a future history that embraced all the peoples of the planet and, during the last quarter of the nineteenth century, some of them reached out to include the inhabitants of other worlds in their visions of the coming universal harmony. Second, the majority of these predictive utopias justified the claim of the eminent Victorian jurist, Sir Henry Maine, that 'the movement of progressive societies has been a movement from Status to Contract', since their regulative principle was the doctrine of constant advance from a less perfect and less fortunate state of society to the more organized and far more prosperous technological world order of the future. Third, the growing sameness of conditions throughout the new industrial societies encouraged the international discussion of these futuristic utopias and the more popular – Bellamy's *Looking Backward* and Hertzka's *Freiland*, for instance – gained an immediate and world-wide following.

Finally, the range of the interests reflected in the new literature was far wider than anything that had ever appeared in the old-style terrestrial utopias, since authors seized on the tale of the future as the ideal means of communicating the most diverse schemes for improving the world. Thus, their proposals varied from the microcosmic society of *Oxford in 1888* to Lamartine's genial vision in 1848 of a prosperous and powerful Ibergallitalian Federation that would enable the French to oppose all British attempts to control Europe. The Oxford prophet of 1838, Richard Walker of Magdalen College, limited himself to the formidable task of imagining the utopian regeneration of what *Punch* had called the Half-Way House to Rome. His reasons throw light on the motivation of utopian fiction in the last century. He begins with the proposition that

> ... the floodgates of knowledge are thrown open by the wide spread of education, through all ranks of society, and this ancient University, almost in spite of herself, is being carried down the stream of change. The struggle of reason and prescription is begun, and if we would exist as a University, to be respected in the sight of England and Europe, we must

> partake in the general superior cultivation of reason, and yield up not a little of our prescriptive rights of thinking and acting.[4]

One unexpected change is the ending of the old self-centredness, since 'the place is no longer considered as almost an exclusive school for orthodox Church of England Divinity, but holds out rewards for various kinds of knowledge in the great scale of truth'. An even more remarkable transformation appears in the report that the old fear 'of the high cultivation of the reasoning powers has quite subsided, and Oxford is holding a place among those Universities, which are willing to admit the converging rays of science and probable knowledge in its concentrated focus, by no means excluding orthodox theology and sound ethics. Religious and moral, physical and intellectual education, each holds a proper place in her improved system.'[5]

It is pleasant to record that later developments proved the prophet had been right in some matters – the Dissenters were admitted to the university; an adequate heating system was installed in the Bodleian Library; and the viva voce system was abolished in favour of written examinations. But there can only be cause for mild regret that the University of Oxford has never been willing to introduce that most desirable improvement whereby every college has an established post for 'a *Conservator of Morals*, whose office it is *strictly* to repress all vice and expense'.

The contrast between the precise social and political theories of Lamartine in *France and England* and the sturdy pragmatism of *Oxford in 1888* was typical of a division that ran through the utopian fiction of the nineteenth century. In general, French writers tended to produce ideal states of the future that began from a theoretical view of the best social system, whereas propagandists in the United Kingdom and the United States favoured a less doctrinaire and more practical method of dealing with the great imponderables of social justice, human equality and the growing complexity of a technological society. The appearance of so many ideas borrowed from Fourier, Saint-Simon and Comte shows that the French utopias were advancing along an axis that ran from the debates of the Enlightenment about man and society to nineteenth-century theories about the application of scientific principles to the systematic organization of society. There are, for instance, clear Saint-Simonian ideas behind Lamartine's dictum that, 'as it is useful to mankind at large that constantly increasing agglomerations should be formed, they will be formed wherever possible, and whenever their necessity is perceptible. Telegraphs, railroads, steamboats, diminishing distance more and more, are every day rendering the extensions of the most prosperous societies more facile and more durable by the spontaneous adjunction of contiguous populations.'[6] Lamartine continues with a passage that echoes the words of Saint-Simon in the *Reorganization of the European Community* in 1814, when Saint-Simon wrote that there 'will come a time, without doubt, when all the people of Europe will feel that questions of common interest must be dealt with before coming down to national interests ... The Golden Age of the human race is not behind us but before us; it lies in the perfection of the social order. Our ancestors never saw it; our children will one day arrive there; it is for us to clear the way.'[7] The Lamartine variation runs:

> The actual condition of Europe proclaims that one of those supreme moments is at hand, when her balance ought to be established on the most extended, the most rational, and by consequence the most solid base. All enlightened men have vague presentiments of this advent; each nation is sensible that all is provisional and precarious with herself and her neighbours. All should make preparation for these changes by a search after the laws they should work upon, in order to be ready beforehand, and to keep pace with the march of events as they are developed; for the happiest, the most prosperous nations will be those which are best grounded in the wants of human nature and best provided for the future.[8]

As Lamartine enlarged on these topics, the politician got the better of the poet and he gave a national and political bias to Saint-Simon's proposals for a united community of the European nations. Lamartine ignored the Saint-Simonian call for a Franco-British agreement as the prerequisite for European unity, and he chose to present his future federation of France,

Belgium, Italy, Portugal and Spain in implacable opposition to a destitute but still perfidious Albion. 'What has become then of this Great Britain', the question goes, 'that has so often upset Europe to maintain her commercial monopoly ... ?' And from the future echo gives the pleasing answer that 'she continues to work out her destiny, and to undergo the fate of all exclusively commercial people'. Shorn of her colonies, banned from the markets of Europe and opposed by the United States, the United Kingdom has gone the way of Tyre and Sidon. *Les lauriers sont coupés.*

This vision of a world designed to the specifications of the École Polytechnique did not please all readers; and two years after Lamartine's projection of the Saint-Simonian doctrines a challenge came from his contemporary Émile Souvestre, who wrote about the unhappy consequences of material progress in *Le Monde tel qu'il sera.* This conflict between the progressive utopia and the regressive dystopia marks the beginning of a dispute that continued in muted fashion into the twentieth century and then became the main theme of prophetic literature after the First World War. For the first time a writer had turned to the tale of the future in order to show that man cannot live by steel and steam alone and that the theories of Robert Owen, Fourier and Saint-Simon arc of themselves incapable of bringing about a better social order.

The method of the book is quite simple – the central character is a young man in search of the world's happiness. He reads the utopian philosophers of the day and discovers he has faith in the future, 'that promised land of all those who cannot see clearly in the present. He believes in the infinite progress of the human race as ardently as the provincial who has been proclaimed a man of letters, and so believes in his career as a writer.'[9] He gains his wish to see the future and arrives in the brave new world of the year 3000 to discover, of course, that technological progress does not bring in the millennium. The rigorous logic of Souvestre's analysis demonstrates with frequent and urgent irony that the blessings of technology will destroy the quality of life if they are not regulated and made to serve the true needs of human beings. It is no surprise to find that the slogan of the fourth millennium is *chacun pour soi*. In the world Republic of United Interests all social, industrial and political activities conform to the doctrines of utilitarianism, and the principle of the division of labour has reached maximum application in the policy that requires each nation to concentrate on one industrial product.

It is interesting to note how Souvestre applies the twin techniques of prediction and propaganda that were to determine narrative form in the twentieth-century brave new worlds. Thus, he is careful to combine the alluring method of describing expected advances in technology with the nay-saying fervour of the earnest moralizer. Souvestre aims to convince the reader with an account of the changed world of the future – planned cities, machines of the most advanced kind, steam submarines, rapid underground transportation. The novelty of these advances serves to confirm the message of values and principles. The lesson begins when the time-traveller suffers his first shock on discovering the abolition of the normal a nd natural (there are steam-feeding systems for all infants) and it ends when the future Candide realizes that the price to be paid for the all-out pursuit of material prosperity is a society of 'perfect machines and brutalized workers'. The last paragraphs sum up the major points of the book by inviting the reader to weep 'for a world in which man had become the slave of the machine and self-interest had replaced love'.

The appearance of the first anti-utopia in the course of futuristic fiction marks the opening of the now painfully familiar argument about the advantages of technological progress. This is to say that in the decade following on the first railways and the first steamships two able writers had used the tale of the future as a means for debating the great question of machinery and mankind. By the 1840s it had become obvious that Europe was going through a process of very rapid change and that the most apparent symptoms of this were the continuing flow of new inventions and the increasing concentration of labour in the great cities. It was a matter for general congratulation and some concern. The themes of *Le Monde tel qu'il sera* were repeated by Dickens a decade later in the harsh industrial context of *Hard Times*; and Lamartine's delight in technological advances had precedents and parallels in Thomas Carlyle, for whom the transformation of European life had an almost magical quality: 'The shuttle drops from the fingers of the weaver, and falls into iron fingers that ply it faster. The sailor furls his sail, and lays down his oar; and bids a strong, unwearied servant, on vaporous wings, bear him through the waters.'[10] There, in 'Signs of the Times', Carlyle tried to reconcile technological progress with the moral sense. He hoped for a balance between those who sought to develop the inner power of the human being and those who dedicated themselves to the outward pursuit of technological improvement; for he feared that the 'undue cultivation of the outward, again, though less immediately prejudicial, and even for the time productive of many palpable benefits, must in the long-run, by destroying Moral Force, which is the parent of all other Force, prove not less certainly, and perhaps still more hopelessly, pernicious. This, we take it, is the grand characteristic of our age.'[11] This recognition of the inescapable struggle between material progress and moral values begins to appear with increasing frequency in the tale of the future from the 1840s onwards. The differences between Lamartine and Souvestre like the equally opposed beliefs of Edward Bellamy and William Morris some forty years later mark the extremes of thesis and antithesis in an uncompromising dialectical conflict that moved some way towards, but never decided in favour of, the Souvestre proposition. And here it is necessary to emphasize that the engaging novelties of these brave new worlds tend to mask the primary fact that all didactic stories about the future begin from the possibilities considered to be latent in the contemporary world. The archaeology of the genre reveals that, level by level and period after period, the perennial question about the best arrangement of human society is at the centre of this intensely self-conscious and exceptionally varied form of fiction. The propaganda of the utopias and the dystopias, the idealized politics behind the forecasts of coming wars, the delight in the wonder-working inventions of science fiction, the rejection of industrial society in the arcadias of the future – all these alternations between the congratulatory and the condemnatory show that the tale of the future is much more than a convenient form of fiction to be used for any purpose; it is the register of moods in a changing society and a running commentary on the search for final solutions to all human problems.

The utopian elements in the science fiction stories of Jules Verne speak for themselves; and even in the seemingly unlikely field of the imaginary wars of the future the same obsessive desire to reach a more perfect level of existence is constantly at work. Most of the hundreds of imaginary wars that appeared between 1871 and 1914 presented idealized visions of the heroic nation triumphant over all enemies and set for ever on a course to increasing power and prosperity. Moreover, even on those occasions when authors chose to describe the defeat of their nations, the forecasts of the coming disaster fell into the pattern of the admonitory terrors of the dystopias, since the propagandists sought to achieve their objectives of political or military reform by demonstrating the catastrophic consequences of wrong principles and chronic national failings. For better or for worse, in war or in peace and in hope or in fear, the authors of all these stories about the future respond to the problems and possibilities of their society. They show how the industrial community should be organized, or how they would change the world for the better, or how the sciences could affect the conditions of existence throughout our world.

The powers of science and the ideal order of industrial society were the great engines of futuristic fiction in the last century. No matter what the differences in history and tradition may have been between the industrial nations, a common experience of the new technologies and a comparable sense of purpose worked impartially on both sides of the Atlantic to produce imaginary wars, scientific romances, technological utopias and ideal states that displayed a frequent similarity in their ideas and styles of writing. In an American utopia of 1836, *Three Hundred Years Hence* by Mary Griffith, the sleeper wakes again to life in a more perfect world that is not noticeably different from another ideal society of 1871, *The Next Generation* by John Francis Maguire, the Member of Parliament for the city of Cork. The American writer looks forward to locomotive cars, dirigible balloons and self-propelling agricultural machines; the Irishman looks forward to steam balloons, a Channel Tunnel completed and similar technological triumphs. Both of them agree that women will obtain political equality with men and that they will have a major role in any society of the future. In the British parliament of 1891, for example, there are 131 women and 399 men; the beautiful Selina Bates is Chancellor of the Exchequer, Meliora Temple is Commissioner of Works, and Grace O'Donnell is the Patronage Secretary. In the American variant on this theme the business women of the year 2135 are as good as the men for the reason that, 'when women were trained to the comprehension of mercantile operations, and were taught how to dispose of money, their whole character underwent a change, and with this accession of business talent came the respect from men for those who had a capacity for the conducting of business affairs'.[12]

This proof by prediction has always been a distinctive feature of the tale of the future; and it is most apparent in the stories of the war-to-come and in the prophecies of imminent social calamities, because these concern themselves with whatever their authors consider to be the most urgent practical problems of their time. So, Edmund Ruffin produced *Anticipations of the Future* in 1860, a propaganda story in which he described the causes and the course of an American civil war ten months before the Confederate attack on Fort Sumter began the real conflict in 1861. Ruffin was a person of consequence in Virginia and South Carolina, where he had done much to improve the standards of agriculture; and, as he explained in a lengthy preface to his Southern readers, he had chosen 'to present his general propositions and argument in a novel, and therefore a more impressive form, by using illustrations and examples of supposed actual occurrences, as the consequences of previous conditions'. His hypothetical history demonstrates the justice of the Southern cause, and in the manner of so many of these projections an ideal South emerges from a just war. Because the Southern states refuse to accept any form of anti-slavery legislation, they secede from the Union and establish a provisional government; and then the imaginary fighting starts after the first belligerent act of the Southern Confederacy—the seizure of Fort Sumter during the night of December 24th, 1867.

In the civil war that followed ...

> ... the northern government, as early as it could act, began and continued to make strenuous efforts to subdue the seceding states, speedily and effectually. Every means of war, which had previously been supposed available and likely to be effective – blockade, invasion, and servile insurrections attempted with preparations and accessories that had been, before trial, deemed the most formidable, mischievous, and certain of effect – all have utterly failed to produce any important effect, in weakening or humbling the seceded states; and all these effects have produced much more loss and damage to the aggressive than to the assailed party.[13]

The war ends a year later – the South a victorious and separate nation, the Northern states ruined and in total disarray.

> As to the predaceous and troublesome New England states with their pestilent fanaticism, no political community or power will be willing to accept their annexation, by union or allegiance. Greedy as England has always been for territorial acquisition and extended dominion, and anxious to retain even the most costly and unprofitable colonial possessions – even England would now refuse to receive, as a free gift, the voluntary annexation of New England to British America.[14]

Although the uncompromising and aggressive ideologies of these imaginary wars may seem to separate them from the more constructive schemes of the ideal states, they were nevertheless a part of the search for the ideal community – united, stable, truly happy and autonomous – that dominated the futuristic fiction produced in Europe and the United States up to the outbreak of the First World War. The idealized chauvinism of the imaginary wars was the most extreme expression of the world-creating activities of predictive fiction before the great divide of 1914; for these stories of the future displayed a self-regarding and self-assured habit of mind that worked with equal effect both in forecasting the best of futures for the best of nations and in describing the greater felicity of life in the coming centuries. The cardinal principle, common to the tales

15 The immense capacity of the applied sciences displayed on the title-page of a book by Jules Verne.

16 In the wrong hands, Verne suggests, science can prove immensely destructive.

17 To the all-powerful sciences anything was possible – even floating palaces in the clouds.

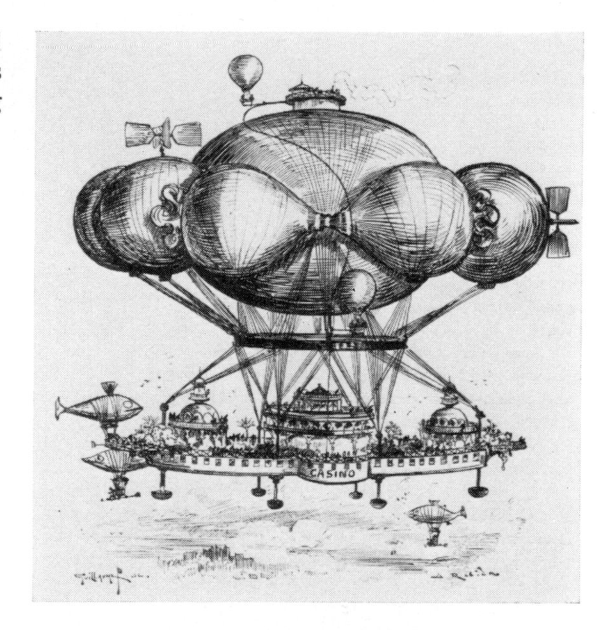

18 Flying machines of the future from a book by Verne's contemporary, Louis Jacolliot, *Les Mangeurs de feu*, 1887.

A. ROBIDA
RÉDACTEUR EN CHEF

La Caricature

JOURNAL
HEBDOMADAIRE

Abonnements d'un an, Paris et départements : 20 francs. — Union postale : 24 francs — Trois mois : 7 francs. — Bureaux : 7, rue du Croissant.

LA GUERRE AU VINGTIÈME SIÈCLE

A. Robida

19 Armoured locomotive-fortresses, *les blockhaus roulants*, go into action in this genial anticipation of warfare in the twentieth century.

20–21 The idea of the future was a favourite obsession with the French artist Albert Robida. His fertile imagination foresaw submarines – and a family gathered round their television set.

22 Robida's method was essentially to domesticate the possible by imagining the uses for scientific inventiveness.

23–24 In like manner the English writer W. D. Hay described the final conquest of nature in *Three Hundred Years Hence*, 1881. The most significant indication of that conquest was the construction of great submarine cities.

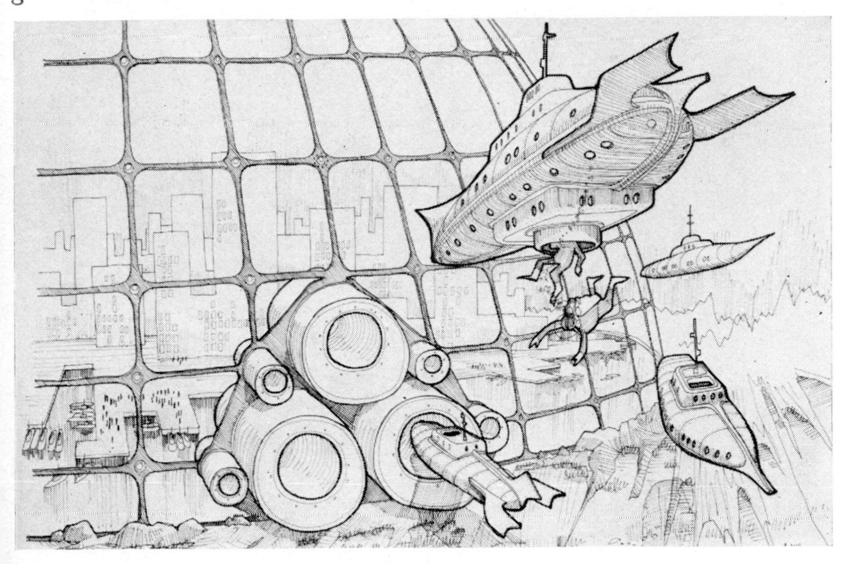

25 With Verne and with most of his contemporaries the power of science encouraged fantasies of vast machines.

of future wars and to schemes for the ideal state, was the assumption that the world would belong eternally to the then dominant technological societies and that all people on Earth would be for ever content to accept the supremacy of the Western nations, for ever happy to adopt Western ideas on government, social organization and cultural values.

To look backward from the end of the century at these developments in the tale of the future is to see that the modern phase in this literature opens in the crucial decade of the 1860s, when tales of the future became more numerous and began to reach their first popular audiences. In that decade new sociological and evolutionary ideas had spread rapidly throughout Europe and the United States. At the same time, as the rate of technological application accelerated, the new histories of the nations and the many popular accounts of social improvements and industrial advances provided a coherent and complete explanation of the past, present and future of the human race which was in total accord with the general experience of material progress. These innovations penetrated the tale of the future with great rapidity – changing the style of the science fiction stories, encouraging the forecasting of future wars, widening the technological base of the ideal states, providing new material for the dystopias, introducing evolutionary scenarios about possible changes of direction in the onward march of mankind. Before the 1860s science had waited on fiction in the tale of the future, because most writers were content to apply the scientific ideas of their time to the subsidiary task of establishing a natural setting for their stories of time-to-come. And then, after Verne had shown what could be done with scientific information in his very successful story *From the Earth to the Moon*, the new fiction began to emerge from the innumerable possibilities contained in contemporary science. That is, as scientific ideas began to decide the shape and substance of many entertaining stories about the limitless powers of technological man, these new stories inaugurated the first major period in the history of science fiction.

There were similar developments in the technological utopias, since the crude notion of constant material betterment encouraged happy visions of future epochs in which the applied sciences would vastly increase the prosperity and well-being of mankind. These most optimistic of utopias were typical of the times. They were a departure from the tradition of the ideal states, since they had little–sometimes nothing at all–to say on the great questions of social justice and the right ordering of human society. In these comfortable prophecies the Gross National Profit was king; and they were popular with readers everywhere, because their forecasts reinforced the confidence of the age. One indication of their popularity was the publication in 1862 of a four-page projection of a twentieth-century newspaper, *The Times, No. 55,567, 1962,* in which posterity was shown to rejoice in the delights of highly advanced technologies. Another indication of the interest taken in these predictions throughout the industrial countries was the printing history of a Dutch forecast of 1865, *Anno 2065* by Pieter Harting. This came out in a German translation in 1866 as *Anno 2066*, then in French in 1870 as *Anno 2070*, and then in English in 1871 as *Anno Domini 2071*. European readers were clearly eager to have the answers to the opening questions: 'When we compare the present condition of society with that of past centuries, the question naturally arises–what will the future be? ... Where are we to look for the fruits of those innumerable germs which the present generation is sowing for the benefit of those that will come after them?' Harting repeated the most cherished beliefs of his time, using the more auspicious advances of the twenty-first century '... to illustrate the fact that the most important discoveries–such as have been most universally brought to bear upon the joint social condition of mankind–have simply resulted from the inventions of scientific men who never dreamt of the practical application of their discoveries'. So, Harting shows that in the future all the most desirable advances will continue to depend upon the great chain of free trade, private enterprise and industrial innovation. Chapter by chapter the author realizes the fondest aspirations of the age: a Channel bridge, a transcontinental railway from Pekin to Berlin, colour photography, dirigible balloons, world-wide telephonic communications, universal literacy and compulsory education in Europe, and a new source of energy in the unexplained form of force-holders and energy-preservers. Best of all, the twenty-first century is to be an era of peace, thanks to the deterrent effects of the new weapons:

> When, in the last war, now about a century ago, the navies of England, France, Russia, and America had mutually destroyed one another; when, through a bombardment from both sides of the Channel, the capitals of England and France had simultaneously been set on fire; when the losses on both sides had become incalculable, not to say irreparable; then, but not until then, people began to ask themselves whether even a victory was worth such enormous sacrifices. And it finally dawned on the public mind that *in all wars the conqueror is likewise the loser.*[15]

The italics presented an early and clear formulation of a belief that became a commonplace in the ideal states of the future after 1871 – the proposition that the more destructive weapons became, the more difficult it would be to wage wars. This conviction depended on many assumptions, and one of the most characteristic of them was the expectation, repeated so often in the literature of progress, that every advance in technology would bring men that much closer to their fellow men. Here is the view of the future as it appeared to Pieter Harting in 1865:

> But that which has mainly contributed to render war gradually a matter of rare occurrence, and which, we trust, will ultimately lead to its complete abolition is the vastly increased intercourse between the peoples of various nationalities, by which all those silly inherited national antipathies have slowly become absorbed; then again, we have the application of the principles of free trade, the removal of all those barriers that separated nations from nations, an universal system of coinage and weights and measures, an increase in the means of locomotion and communication, and the fusion of the individual interests of particular nations into one great universal 'public weal'. Nations have ceased to stand opposite, against one another; they flourish side by side; by thousands and thousands of bonds they are joined and held together; and if the nineteenth century has witnessed the introduction of the principle of nationality, ours

has made another step in the right direction, and produced the recognition of the principle of humanism.[16]

This Dutch utopia and the evidence of the contemporary prophecies in France confirm the claim that science and industrialism were the decisive factors in the final evolution of futuristic fiction in the 1860s. In 1862 Jules Verne produced the first of the new-style science fiction stories; and within the elaborate calculations and high adventures of *Five Weeks in a Balloon* he buried, prophecy within prophecy, two alternative visions of the future. As the *Victoria* speeds across the fertile plains of Mfuto, the optimistic engineer Samuel Fergusson foresees the day when Africa will be the heart-land of a new world: 'The nations of the future may come here when the land in Europe is no longer able to feed the inhabitants ... and this world over which we are soaring, more fertile, richer, more virile than the rest, will become some great kingdom where discoveries will be made greater even than steam and electricity.' Was it deliberate contrivance that made Verne present the contrary opinion through the person of that open and candid sportsman, Dick Kennedy? That canny Scot feared that 'the time when industry gets a grip of everything and uses it to its own advantage may not be particularly amusing. If men go on inventing machinery, they'll end up by being swallowed up by their own machines. I've always thought that the last day will be brought about by some colossal boiler heated to three thousand atmospheres blowing up the world.' The final word came from the faithful servant, the rare and honest Joe, who said: 'I bet the Yankees will have a hand in it.'

As Jules Verne started on the first epic prophecies of the first industrial age, some of his contemporaries continued the debate about the best kind of future that Lamartine and Souvestre had begun: Jacques Fabien, *Paris en songe*, 1863; Victor Fournel, *Paris nouveau et Paris futur*, 1865; H. Mettais, *L'An 5865*, 1865; Ernest Jonchère, *Clovis Bourbon: excursion dans le vingtième siècle*, 1868; Fernand Giraudeau, *La Cité nouvelle*, 1868; Tony Moilin, *Paris en l'an 2000*, 1869. These present the established range of possible futures: from the simple account Fournel gives of the admirably planned Paris of 1965 to the customary horrors of an unfeeling materialism in the future world Giraudeau describes; from the socialist propaganda in the ideal Paris of Moilin to the punishments Mettais inflicts upon the French for their sins. The rambling account by Mettais in his tour of the world in the year 5865 is an early example of the prophetic histories of the natural catastrophes that wipe out contemporary civilization for the best of moral reasons; and it is yet another indication that the tale of the future had been accepted as a natural means of propaganda by the 1860s. In the preface Mettais sets out the failings of his day: 'When I see so many vices spreading through every part of our society and I note that the worst of them is selfishness, then I am afraid for our society.' From that everyday piece of morality Mettais goes on to make his protest, very unusual at that time, against the spread of European colonialism: 'What is the meaning of all these battles that are going on from one end of the world to another; these arrogant demands of the strong upon the weak; these endless massacres; this pillaging by armed force?'[17] For Mettais these are 'the symptoms of the dissolution of society', and in his imagination he employs nature to inflict a just retribution. In the distant future France has vanished and Paris is a collection of miserable huts. Volcanic eruptions have destroyed Western Europe; the achievements of nineteenth-century industrialism have been long forgotten in the sixth millennium when a Cossack empire rules in Russia and the former colonial peoples of the world have grown into powerful nations. The barbarians of the future ask the obvious question: 'How has it come about that the powerful France of the past has fallen so low that it is no more than a miserable tribe lost in the deserts of barbarism, reduced to a few thousand individuals?' And history answers that 'when a nation is far gone in the love of luxury, when money serves for nothing more than a means to pleasure, then that nation is about to start on the road to ruin.'

> Religion blushing veils her sacred fires,
> And unawares morality expires.
> Nor public flame, nor private, dares to shine;
> Nor human spark is left, nor glimpse divine!

> Lo! thy dread empire, Chaos! is restor'd;
> Light dies before thy uncreating word;
> Thy hand, great Anarch! lets the curtain fall,
> And universal darkness buries all.[18]

These calamitous prophecies had their origin in the long debate about man and machinery – the citizen in the new society – which had begun in the first quarter of the century. As the great cities spread across the countryside, as the urban populations continued to multiply and as an explosion of inventions changed the world, the tales of the future had grown in number and popularity until they burst upon the Europe of 1871 in a tidal wave of forecasts, imaginary wars, utopias, dystopias, and technological fantasies.

One reason was the rapid advance in armaments during the 1860s – the Whitehead torpedo, the all-iron warship, breech-loading artillery, the Chassepot rifle, the Gatling gun; and when these were added to the evidence of new methods of fighting in the Austro-Prussian War, the American Civil War and the War of 1870, there could only be one conclusion. The wars of the future would be different. And so in 1871 Chesney wrote his story of a successful German invasion of the United Kingdom, and the *Battle of Dorking* became the model for the forecasts of future wars that kept the presses turning throughout Europe. Another reason lay in the equally convincing demonstrations of technological progress in the opening of the Suez Canal, the laying of the Atlantic Cable, the first transcontinental railway in the United States, and the beginnings of pasteurization. These advances strengthened the general belief that the scientists could and would find an answer to every problem – a belief that was repeated with total conviction by the editor of the middle-class illustrated magazine, *The Graphic*. He made a special point of saying in the issue for June 3rd, 1871, that

> ... a young surgeon anxious to distinguish himself as a scientific enquirer could not perhaps take up any subject more promising than that of hydrophobia ... It is really time that medical men inquired into the terrible subject in something like a scientific method. Will any competent authority, for instance, subject the saliva of a mad dog to chemical

analysis, and tell us what is the nature of the poisonous element which is capable of circulating through the blood and affecting human beings with such distressing symptoms?

In this way the sciences raised the level of hope, encouraging expectations of even greater advances that found their first popular expression in the science fiction of Jules Verne and in the technological utopias of the 1870s: in J. F. Maguire's *The Next Generation*, 1871; Cyrus Elder's *Dream of a Free-trade Paradise*, 1872; Jókai Mór's *Jövó Század Regénye*, 1872; E. Maitland's *By and By*, 1873; A. Blair's *Annals of the Twenty-Ninth Century*, 1874; in the anonymous, *In the Future*, 1875; in Alphonse Brown's *La Conquête de l'air*, 1875; Benjamin Richardson's *Hygeia – a City of Health*, 1876; Georges Pellerin's *Le Monde dans deux mille ans*, 1878; Jules Verne's *The Begum's Fortune*, 1879; Kurd Lasswitz's *Bilder aus der Zukunft*, 1879. These visions of the future often ran to many editions and they were translated into foreign languages, because they were books of revelation for an epoch of incessant change.

As Chesney had demonstrated to the whole world, the greatest power of these predictions lay in their capacity to make a great issue come alive in the dramatic circumstances of an imagined future; for it has always been a condition of futuristic fiction that, whenever the strongest inclinations of a writer coincide with the mood of the moment, there is an immediate and widespread interest in whatever is said about the future of the world or the ordering of society. In 1871, for instance, Bulwer-Lytton's famous story of *The Coming Race* went through five editions and was considered important enough for the *Annual Register* to hand it down to posterity as 'an ingenious book ... which described an ideal world, supposed to represent the state of things towards which we are gradually tending'.[19] The evidence for this intimate association between the image of the future and the expectations of society finds additional support in the observations of James Presley, Director of Cheltenham Library. In 1873 Presley drew attention to the rapid increase in the number of utopian stories: 'The last four or five years', he pointed out, had been 'remarkably fruitful in works of a Utopian character ... No doubt this is due to the stimulus derived from two circumstances: the increasing attention paid of late years to the study of social science; and, secondly, to the new political influences resulting from the late Franco-German war.'[20] Again, the myth-making activities of utopian fiction found full recognition from John MacNie, a graduate of Glasgow University who emigrated to the United States and became the first professor of mathematics in the University of North Dakota. In the preface he wrote in 1889 for the second edition of his ideal state, *Looking Forward, or the Diothas*, MacNie had this to say about the tale of the future in his day:

> It is not strange that the present age is so fruitful in works that anticipate the future of our race. The tendency is rather to look forward with hope than to admire the past. The extraordinary advances made during the past century in science and mechanical invention have naturally raised hopes of a corresponding advance during the coming centuries and stimulated the impulse to forecast the hoped-for glorious future. Such a forecast need not necessarily be a mere exercise of fancy, adapted to amuse an idle hour. It may serve a useful purpose, both in showing by contrast the evils of our present social organization and by acting as an incentive for each to do his best towards the attainment of a loftier social ideal. The aim of the present work is to give such a forecast of the future of our race as may be inferred with some reasonable probability from present conditions and tendencies. Far from indulging in mere flights of the imagination, the author has earnestly endeavoured to keep within the bounds of sober reason. Some of the changes or events suggested are such as are not unlikely to result from tendencies now in operation. Others, again, are intended to foreshadow changes now impossible to predict but of corresponding importance.[21]

Although the observations of Presley and MacNie point to some of the major factors and more immediate developments responsible for the sudden outpouring of futuristic stories that began in 1871, a full explanation would show how this remarkable advance in the power and popularity of the tale of the future was the primary result of dominant ideas about mankind, industrialism and the future of society that had come together for the first time in the 1860s. The oldest and most potent of these was the idea of progress which had animated utopian fiction from the start in Sebastien Mercier's *L'An 2440*. When this was fused – or sometimes confused – with the Darwinian ideas that became general knowledge after the publication of *The Origin of Species* in 1859, the revised doctrine of progress provided new matter for tales of the future at the same time as it seemed to add the sanction of evolutionary theory to the description of life in the centuries ahead.

Before the expansion of utopian fiction about 1871 the argument in the ideal states of the future had advanced mechanically like an engine on a single-track railway, rolling forward from the blessing of social and industrial progress to the revelation of the far better scheme of things in another age. And then, once the Darwinian theory of natural selection and the struggle for existence had demonstrated the succession and mutability of all living things, the idea of evolutionary development became an article of faith in the sacred literature of progress. In 1867 the formidable apostle of Darwinism in Germany, Ernst Haeckel, proclaimed *ex cathedra* in the University of Jena that the evolutionary theory had explained the past and illuminated the future:

> Our Theory of Development explains the origin of man and the course of his historical development in the only natural manner. We see in his gradually ascensive development out of the lower vertebrata the greatest triumph of humanity over the whole of the rest of Nature. We are proud of having so immensely outstripped our lower animal ancestors, and derive from it the consoling assurance that in future also, mankind, as a whole, will follow the glorious career of progressive development, and attain a still higher degree of mental perfection. When viewed in this light, the Theory of Descent as applied to man opens up the most encouraging prospects for the future.[22]

The assurances of a general advance from good to better, and from generation to generation, fixed the point from which the progressive ideal states of the future began. For those who abhorred the delights of material improvement there was the contrary model of the regressive utopia, which showed that a single day in arcadia is better than a thousand years in Technopolis. And there were those dark dystopian projections in which the disdainful and the indignant carried the enthusiasms of the age to the worst possible conclusions.

6

The Best of Worlds and the Worst of Worlds

Edward George Bulwer-Lytton, first Baron Lytton of Knebworth, did not share the contemporary enthusiasm for the future of mankind. In the course of a long life he had observed with increasing disillusionment all the great advances of the age; and in 1870 he began work on *The Coming Race*, into which he wrote his feelings of dismay at the exaggerated moral pretensions that so often went with the idea of material progress.

The book was based on the first major application of evolutionary ideas in the history of fiction. As Bulwer-Lytton explained in a letter to his friend, John Forster, the scheme of the story depended on 'the Darwinian proposition that a coming race is destined to supplant our races, that such a race would be very gradually formed, and be indeed a new species developing itself out of our old one, and that this process would be invisible to our eyes, and therefore in some region unknown to us.'[1] Deep below the surface of the earth the evolutionary process had created the highly advanced race of the Vril-ya whose social organization and mechanical appliances made the Victorian world look like Darkest Africa. The achievements of the coming race represented the denial of contemporary hopes for the improvement of mankind. Thus, John Stuart Mill's argument for the equality of women was pushed to a fearful conclusion in the description of the Gy-ei, who are superior to their men in intelligence and physical power. The brotherhood of man, preached by the Christian Socialists, found its fulfilment in an egalitarian society in which there is 'nothing to induce any of its members to covet the cares of office'; and the idea of progress was ridiculed by the picture of a highly mechanized civilization which is unable to solve the problem of leisure.

As Bulwer-Lytton develops the conventional themes of the satirist, he displays the most comprehensive use of scientific material to appear in the English fiction of his day. There are, of course, all the mechanical marvels that create the precise technological setting for the story: automata, aerial vehicles, musical machines, sleep-teaching, and an unprecedented source of power in the fluid *Vril*, 'the mightiest agency over all forms of matter, animate or inanimate'. Of far greater significance, however, is the original and competent way in which Bulwer-Lytton draws on scientific information to create the illusion of a different race and another world. He ascribes the superiority of the Vril-ya to 'the intensity of their earlier struggles against obstacles in nature amidst the localities in which they had first settled'. He traces the shape of their heads to 'a development, in the course of countless ages, of the Brachycephalic type of the Age of Stone in Lyell's *Elements of Geology*'; and he devotes a whole chapter to an account of the Vril-ya language, with sample paradigms and declensions, and he refers to the linguistic theories of Max Muller, Taylorian Professor of modern languages in the University of Oxford. After such a display of contemporary knowledge it is not surprising that Bulwer-Lytton should send his son news of a favourable reception of the book by learned and literary readers. 'Blackwood tells me', he wrote, 'that the opinions he hears privately are very enthusiastic, chiefly from professors and scholars.'

The book appeared on May 1st, 1871, the day when Blackwood's of Edinburgh also published Chesney's *Battle of Dorking* in their magazine, and by an even more extraordinary coincidence the day when Samuel Butler brought the manuscript of *Erewhon* to Chapman and Hall. Moreover, in unconscious imitation of *The Coming Race*, Butler had chosen to present his book as the account of a highly advanced race at a certain stage in its evolution; but unlike Bulwer-Lytton his intention was to attack the improper treatment of Darwinian ideas by driving home, as he said, the specious misuse of analogy.

He made his point in a famous passage:

> I learnt that about four hundred years previously, the state of mechanical knowledge was far beyond our own, and was advancing with prodigious rapidity, until one of the most learned professors of hypothetics wrote an extraordinary book ... proving that the machines were ultimately destined to supplant the race of man, and to become instinct with a vitality as different from, and superior to, that of animals, as animal to vegetable life. So convincing was his reasoning, or unreasoning, to this effect, that he carried the country with him; and they made a clean sweep of all machinery that had not been in use for more than two hundred and seventy-one years.[2]

That calculated application of Darwinian ideas and the construction of a divergent evolutionary history for the Vril-ya were the first signs of a major shift in the pattern of expectation. It meant a sudden enlargement of the imagination within the tale of the future; for the great virtue of the Lytton–Butler Effect lay in the dramatic use the two authors had found for the theory of evolution. In consequence, the myriad possibilities for change in man and nature have ever since been a staple of futuristic fiction. Once the concept of biological change was embedded in the idea of the future, writers turned Darwinism into a system of rewards and punishments for the instruction of their readers. Thus, Darwinism proved an infallible means of showing that in the nature of things the world must go from bad to worse, as some earnest authors demonstrated in the admonitory prophecies of H. C. M. Watson in *Erchomenon; or, the republic of materialism* (1879), A. M. Brookfield in *Simiocracy* (1884), W. Grove in *The Wreck of a World* (1889) and J. A. Mitchell in *The Last American* (1889).

The most extensive and most effective application of Darwinism, however, first appeared in *The Time Machine* in 1895, when H. G. Wells pursued the nineteenth-century striving for the control of nature to a grim, evolutionary conclusion. In the year 802,701 the Time-Traveller discovers the final irony of human history. Because men had been able to eliminate disease and clear the world of all harmful things, the machinery of the struggle-for-existence had inflicted nature's revenge upon the effete remnants of a decadent humanity:

> It seemed to me that I had happened upon humanity upon the wane. The ruddy sunset set me thinking of the sunset of mankind. For the first time I began to realise an odd consequence of the social effort in which we are at present engaged. And yet, come to think of it, it is a logical consequence enough. Strength is the outcome of need: security sets a premium on feebleness. The work of ameliorating the conditions of life – the true civilising process that makes life more and more secure – had gone steadily on to a climax. One triumph of a united humanity over Nature had followed another. Things that are now mere dreams had become projects deliberately put in hand and carried forward. And the harvest was what I saw![3]

That pessimistic rejection of the Victorian dream precedes the even more fearful discovery of the Undergrounders. The Morlocks, who are descended from the labouring classes of a cruel capitalistic society, are nature's instrument for the punishment of mankind: 'Ages ago, thousands of generations ago, man had thrust his brother man out of the ease and the sunshine. And now that brother was coming back – changed! Already the Eloi had begun to learn one old lesson anew. They were becoming re-acquainted with Fear.'[4]

The theory of evolution served other writers as their point of departure for dreams of a happier life in the coming centuries. In the 1880s, for instance, the new theme of the return to nature entered the tale of the future – first, and partially, in the obliteration of industrial civilization which Richard Jefferies so lovingly described in *After London* in 1885; second, and far more effectively, in the forest world of *A Crystal Age* of 1887, in which W. H. Hudson created the prototype for the modern arcadia of the future. The natural catastrophes that return the world to a state of nature draw attention to the changes that had taken place in this fiction since the appearance of *Le Dernier Homme* in 1805. Where Cousin de Grainville had written in the traditional context of a divinely appointed end to the world, Jefferies and Hudson began with the evolutionary disturbances that changed the conditions of human existence throughout the planet. Again, the total obliteration of Victorian society in these two stories was a departure from all the earlier visions of the final calamity. In the disasters described by R. F. Williams in *Eureka* (1837), by Hermann Lang in *The Air Battle* (1859) and by H. O'Neil in *Two Thousand Years Hence* (1868), the authors had concentrated their moralizing into the significant contrasts they discovered between the prosperity of future nations and the desperate, degraded state of the British people. Jefferies and Hudson had chosen the axe in place of the lash; and so, by yet another coincidence in the course of futuristic fiction, another pair of able writers had introduced a new theme in their exploding fantasies of the universal cataclysm.

Jefferies and Hudson had much in common: they were both distinguished naturalists; they had evolved new styles of writing about life in the country; they worked and re-worked the experiences of their youth into their fiction; and they shared a most vigorous detestation of the squalid, dreary, unsatisfying life that had become the norm for the masses in the great new towns. In violent and agreeably anticipative acts of destruction the two men called on nature to wipe out the infamy of urban civilization. Thus, Jefferies discharged his hatred through the natural symbolism of the great lake that had drowned the Midlands of England and the enormous swamp that marked the site of London:

> For this marvellous city, of which such legends are related, was after all only of brick, and when the ivy grew over and trees and shrubs sprang up, and lastly, the waters underneath burst in, the huge metropolis was soon overthrown ... It is a vast stagnant swamp, which no man dare enter, since death would be his inevitable fate. There exhales from this oozy mass so fatal a vapour that no animal can endure it. The black water bears a greenish-brown floating scum, which for ever bubbles up from the putrid mud of the bottom. When the wind collects the miasma, and, as it were, presses it together, it becomes visible as a low cloud which hangs over the place. The cloud does not advance beyond the limits of

> the marsh, seeming to stay there by some constant attraction; and well it is for us that it does not, since at such times when the vapour is thickest, the very wildfowl leave the reeds, and fly from the poison. There are no fishes, neither can eels exist in the mud, nor even newts. It is dead.[5]

A similar hatred of contemporary England runs through the description of the perfect natural world in *A Crystal Age*. In the far-off future great forests cover the whole of Europe and the teeming populations of the nineteenth-century cities have dwindled to small groups of men and women who live a close family-life in a Rousseau's paradise of tame animals, handicrafts and the cheerful cultivation of their gardens. It is evident that the licence to play lord of the universe, granted by the tale of the future, encouraged Hudson to purge his feelings of anger and resentment in an apocalyptic vision of the world renewed. He destroyed everything he disliked in

> ... a sort of mighty Savonarola bonfire, in which most of the things once valued have been consumed to ashes – politics, religions, systems of philosophy, isms and ologies of all descriptions; schools, churches, prisons, poorhouses; stimulants and tobacco; kings and parliaments; cannon with its hostile roar, and pianos that thundered peacefully; history, the press, vice, political economy, money, and a million things more – all consumed like so much worthless hay and stubble. This being so, why am I not overwhelmed at the thought of it? In that feverish, full age – so full, and yet, my God, how empty! – in the wilderness of every man's soul, was not a voice heard crying out, prophesying the end?[6]

Hudson recognized the wish-fulfilment elements in *A Crystal Age*, when he began the preface to the second edition with an admission: 'Romances of the future, however fantastic they may be, have for most of us a perennial if mild interest, since they are born of a very common feeling – a sense of dissatisfaction with the existing order of things, combined with a vague faith in or hope of a better one to come.' He went on to say that 'when we write we do, as the red man thought, impart something of our souls to the paper, and it is probable that if I were to write a new dream of the future it would, though in some respects very different from this, still be a dream and picture of the human race in its forest period.' Although Hudson was familiar with romances of the future, as the preface shows, he does not seem to have realized that in *A Crystal Age* he had introduced the innovation of the apocalyptic arcadia to the tale of the future. His dream of the human race in its forest period displays the permanent mood that informs these rural visions of the good life. By an act of imagination Hudson had returned in spirit to the primordial experience of man in harmony with nature that had first appeared in Greek poetry. It was the substance of the *Idylls* of Theocritus and the inspiration of a poem by Aratus, who anticipated the sentiments of Hudson by more than two thousand years:

> The hurt of strife they knew not in their day,
> Nor yet sharp quarrel and the noise of war.
> Simply they lived, the rude sea far away,
> No ships to bring their living from afar;
> But cows and ploughs and Justice in her rule
> Freely gave all, of just gifts bountiful.[7]

Hudson re-wrote the conventions of the ancient dream in keeping with the evolutionary ideas of his day and the promptings of a profound, millennial impulse to bring time to a halt in the eternal springtime and perpetual perfection of his earthly paradise. So, he transferred the golden land of the classic arcadias from the imagined geographies of Sannazzaro and Sir Philip Sidney to the never-never world of the future. For the implicit suggestion in *A Crystal Age* is that a Wordsworthian law of human regeneration operates throughout nature. Once some catastrophe, similar to the fires and floods of ancient mythology, has removed all traces of a corrupt industrial society, then the survivors will discover the true meaning of human life and will, therefore, move up the evolutionary ladder to a higher level of existence. Fertile fields, gentle animals, pure streams, the clearing in the forest and the dominant role of the Mother in the totally harmonious life of the House—these are the major elements in an archetypal pattern of redemption and renewal that is one of many signs pointing to a growing anxiety about the terms of life in the industrial society of the late nineteenth century.

One response was the dream – the deliberate presentation of a better, happier, nobler way of life. So, William Morris imagined the good life in the *News from Nowhere* and he blotted out the bad life in the legendary tales and chivalrous episodes of *The Earthly Paradise*. In his prologue the dreamer of dreams, as Morris called himself, offered his readers the choice between good and evil:

> Forget six countries overhung with smoke,
> Forget the snorting steam and piston stroke,
> Forget the spreading of the hideous town;
> Think rather of the pack-horse on the down,
> And dream of London, small and white and clean
> The clear Thames bordered by its gardens green.

That vision of an ideal London and, more especially, the return to Eden in the dream sequences of *A Crystal Age* and the *News from Nowhere* are telling indications of the way in which the ancient myths have continued to flourish within the tale of the future. Both Hudson and Morris go through the mythic ritual of destroying their world – changing it out of all recognition – in order that it may be recreated in a more perfect future. The result is total happiness. The time-traveller brings the good news from the future in *A Crystal Age*: 'It seemed to me now that I had never really lived before, so sweet was this new life – so healthy, and free from care and regret. The old life, which I had lived in cities, was less in my thoughts on each succeeding day; it came to me now like the memory of a repulsive dream, which I was only too glad to forget.'[8]

The dream of total harmony in *A Crystal Age* was the first major account of an alternative and radically different kind of society to appear in the tale of the future. It was not a popular book; and it now stands apart from the mainstream of the genre in prophetic anticipation of the rejection of technological progress in favour of arcadian simplicity that was to be a favourite subject with writers between the two world wars. A century ago readers found that the progressive utopias were far more in keeping with their expectations, since they projected the most admired conditions of the age in visions of universal peace, fabulous technological developments, the continued expansion of the industrial nations, even some scheme or other for a world society. In the last decade of the century many of the would-be rulers of the world went forward in imagination to the founding of empires throughout our planetary system. So, Arthur Bennett in his artless and self-confident account of *The Dream of an Englishman* in 1893 described the great federations of the twenty-first century: the United States of Europe, the United States of South America, and the United Empire of Great Britain which gave George III his revenge by including the United States with Australasia, India and Africa. The author ends in the manner of contemporary imperialism by looking for new worlds to conquer. His theme is – wider still and wider: 'And some of the boldest of "the coming race" were looking out upon the stars, and wondering if there were worlds to conquer there. And the federations which their poets sang of, now, were federations of the solar system, and the union which they dreamed embraced Orion and the Pleiades.'[9] The excessive ardour and frequent violence of these imperializing dreams were the product of powerful feelings about the future of the nations that ran through the tales of imaginary warfare and the more popular accounts of the world societies of the future. Here, for example, are the prospects for life in the twenty-third century as R. W. Cole, the author of *The Struggle for Empire*, described them in 1900:

> And so the Anglo-Saxon race went on wresting fresh secrets from Nature every day, while its individual members were continually acquiring more possessions and building more imposing palaces. To know at that time meant to possess. Exulting in their might, the gray-haired scientists steered their vessels through the dark depths of space, while they ransacked worlds for treasures and luxuries; some even towed great masses of valuable rock or precious metal behind their ships. Rare and beautiful plants were uprooted, and strange animals were captured and stowed away in the interior of the ships, and finally deposited in London or the other great cities of the world. Whole families would band

> together and buy an interstellar ship, and rush out into space to seek for themselves a new country and more splendid fortunes.[10]

This mechanical application of imperialistic practices shows how the freedom of futuristic fiction brought out the worst in many writers, who turned their minds to devising the greatest imaginable benefits for their own countries. Some Americans, for example, were convinced that a manifest destiny would oblige the half-brother of the world to take control of the Canadian people. In 1881 the author of *1931: A Glance at the Twentieth Century*, Henry Hartshorne, looked forward to the day when his country would include 'the three great States which once formed the Dominion of Canada, and the outlying territories of Greenland, Labrador, Hawaii, Cuba, and St Domingo.' The proposition came from the debate, then beginning in the 1880s, about the role of the United States in world affairs and the need for an adequate American navy. These ideas ran through the chapter on 'Our Future Foreign Policy' in a forecast of 1888, *Glimpses of the Future*, in which David Goodman Croly argued in favour of an American imperialism: 'We cannot breed a race of great statesmen until we have our say in international matters. We shall meet with some unpleasant surprises when we desire to take our place among the nations of the earth. We are not only without colonies and distant naval stations, but our coasts are surrounded with fortifications in the hands of foreign powers.'[11] The right answer in this game of abstractions is to secure Canada for the United States, because 'there is every human reason why this dependency of Great Britain should become a part of our Union, but the unnatural barriers in the way will probably finally be broken down by force.' The achievement of that dream appeared in the account of the world in A.D. 2000 with which John Jacob Astor, grandson of the first and more notorious of the Astors, began his *Journey in Other Worlds* in 1894:

> The bankrupt suffering of so many European Continental powers had also other results. It enabled the socialists – who have never been able to see beyond themselves – to force their governments into selling their colonies in the Eastern

> hemisphere to England, and their islands in the Western to us, in order to realize upon them. With the addition of Canada to the United States and its loss to the British Empire, the land possessions of the two powers became about equal, our Union being a trifle the larger. All danger of war being removed by the Canadian change, a healthful and friendly competition took its place, the nations competing in their growth on different hemispheres. England easily added large areas in Asia and Africa, while the United States grew as we have seen. The race is still, in a sense, neck-and-neck, and the English-speakers together possess nearly half the globe.[12]

All these visions of the future—ideal wars, ideal states, ideal worlds, the triumph of women and the triumph of nations—were the necessary and normative myths of the first industrial age. Their prophecies were a revelation to their societies and, on many occasions, a call to action in their time. During the last decade of the century the French engineer turned political philosopher, Georges Sorel, had started his meditations on the problems of the age—the illusions of progress, the nature of industrialism, the necessity of violence. And Sorel found that one of the most powerful means of hastening the transformation of any society was 'the framing of a future'. He believed that this would be more effective, 'when the anticipations of the future take the form of those myths, which enclose with them all the strongest inclinations of a people, of a party or of a class, inclinations which recur to the mind with the insistence of instincts in all the circumstances of life.'[13] Although Sorel derided and denounced the fabrication of utopias, and in particular 'the *ridiculousness* of the novels of Bellamy', the course of futuristic fiction in the last century shows that the utopian myths were a major influence in spreading ideas about capitalism, socialism, industrialism, and imperialism. Their special function was to create instructive visions of some future age in which a just, regulated and harmonious world state (or a small and self-sufficient community) had solved the problems of urban expansion, growing populations, increasing friction between classes, the dangers of the arms race, and the many unhappy consequences that had followed on the industrialization of the customary and communal societies of the past. Their greatest power lay in the way the propagandists were able to manipulate the tale of the future by transforming the circumstances of the age so that their histories of the coming centuries demonstrated the predestined advance to the more perfect urban society of the next millennium, whilst their dreams of rural tranquillity satisfied the desire for a truly natural and convivial way of life.

From the start in Bulwer-Lytton's *The Coming Race* and Maguire's *The Next Generation* in 1871 to the ending of the old optimism after the publication of Wells's *The World Set Free* in 1914 all these utopias had maintained an unbroken and uncompromising opposition between the elaborate metropolitan schemes of the kind Edward Bellamy gave to the world in *Looking Backward* and the close, personal relations William Morris described in his *News from Nowhere*. However, no matter how much the authors differed in their conclusions, they all began with the one question that went round the world from Boston to the British Museum – how were the urban masses to organize and conduct their lives? And they found the beginnings of their answers in Charles Darwin, or Karl Marx, or Herbert Spencer, or Henry George, or even in Schopenhauer and Nietzsche. In a sense their contrasting visions of the good society were simply the most recent stage in the long-running debate between reason and nature; but behind all of them and all the time there was the more or less clear realization of the inescapable confrontation between the kinship and neighbourhood relationships of the community and the contractual, individualistic and competitive conditions of capitalistic society. Those were the issues that the great German sociologist Ferdinand Tönnies first presented to the University of Kiel in 1881. Like the authors of so many ideal states, he pointed to the 'mutual possession and enjoyment and also possession of and enjoyment of common goods' that marked the organic community; and he contrasted these conditions with the atomizing forces of urban civilization in which 'every person strives for that which is to his own advantage and affirms the actions of others only in so far and as long as they can further his interest.'[14] These were the terms that decided, in one way or another, the structure and working of all the utopias of the future during their major period of growth and influence between 1871 and 1914; for they all found their natural dimensions somewhere between the polar opposites of the organized super-state and the confident, self-reliant, small community that G. K. Chesterton described with such wit and affection in the hilarious episodes of *The Napoleon of Notting Hill* in 1904.

Many of the architects of the mercantile, industrial state of the future had little to say that had not been said before. They repeated the convictions and platitudes of all who were well pleased with the conditions of their time; and in their most extravagant dreams the best their imaginations could provide was a more advanced version of nineteenth-century society. There was, for example, the Missouri lawyer from Platte City who sang of the glorious future in *A Century Hence and Other Poems*, and in a series of thudding stanzas he foretold how the American dream would come true. A man from the year 1980 speaks of the marvellous advances that show how God had blessed America:

> He told of his visit to Paris and Rome –
> Of his flight over England and Spain;
> He found in his travels no land like his home,
> No place where he wished to remain.
> He spoke of his trip to the banks of the Nile,
> To cities now crumbling to dust –
> Of China, Japan and Australia's isle,
> Where all are in ruin and rust.
>
> He rested on top of the high Rocky Mountains,
> And turned for a view of the West;
> The land was a garden, with forests and fountains –
> A home for the free and the blest.
> He turned to the East, and a picture more bright,
> Never rose in the poet's sweet dream;
> The land was an Eden of love and delight,
> With mountain and valley and stream.
>
> In the midst, at St. Louis, the Capitol loomed,
> With lofty and glittering steeple –

The seat of a Nation, where freedom first bloomed,
Containing a billion of people.
'And now,' he exclaimed, 'the whole Continent's ours,
From Panama, North to the Pole!
For naught but the ocean can fetter our powers,
Or give us less than the whole!'

As we walked to the house, my companion reported
That roads through the land were not found,
That men, on light wings, in the atmosphere sported,
Or walked, as they pleased, on the ground.
With the new motive power, one man could do more
Than fifty, without it, could do;
So people were able to add to their store,
And be generous, noble and true.

An order for supper, by telephone, now,
Had scarcely been made, by my host,
When in sprang a servant, I cannot tell how,
With coffee, ham, biscuit and toast.
He'd come from St. Louis, three hundred miles out,
With dishes delicious and rare;
There were venison, and turkey, and salmon, and trout,
With pine-apple, orange and pear.[15]

The bounteous prospects of a future that would rain good things upon good people was the point where the science fiction marvels of Jules Verne gave place to extraordinary dreams of united and unbelievably prosperous world societies. The most spectacular of these was the work of William Delisle Hay who imagined a future age in *Three Hundred Years Hence* in which political idealism combined with savage Darwinian ideas to establish the all-white and super-efficient Empire of Humanity. Hay outdid Jules Verne in the variety of his technological inventions: flying machines, passenger-carrying submarines, magnetic heat-generators, rust-proof metal alloys, new means of transportation – all the standard devices are present; but of far greater interest is the way in which Hay sought to predict how a great technological society would be the special creation of the most advanced engineering skills. His ideal society was, of course, his understanding of the social ideas and scientific achievements of the Victorian period carried forward to a prodigious perfection. The polemic running through the narrative begins with the Malthusian terror of a vast rise in world population; and Hay rebukes his contemporaries for their lack of faith in science by projecting a conventional Victorian answer into the future: 'Philosophers of three hundred years ago could not have feared numerical increase, had they examined into their present and compared it with their past. They would have seen that the majestic power implanted in Man – his Reason – enabled him to search for and discover the resources of Nature, as the need for them arose.' Science has found all the answers to the problem of a world population of 130,000 millions: great submarine cities collect the mineral wealth of the oceans and the algoculturists harvest fields of sea-food deep beneath the waves; thermal-conductors tap the internal heat of the earth so that bananas grow in Greenland and maize in Iceland; fungus-farms in vast excavated areas within the earth add to the food supply; the prodigious basilico-magnetic boring machine brings water to the barren regions of the world; the engineers find 'means to level down the tops of mountain-ranges, until the table-lands thus formed were of no greater height than was practicable for cultivation.' Finally, the Oecumenic Parliament issued 'The Terrane Exodus Decree' which ordains that 'Cities of the Sea should at once be built in sufficient number and size to contain the population of the world.' The water surface is divided into sixty states into which the existing states are to be merged.

The first step is to destroy all wild and domesticated animals throughout the planet: 'We needed the land and we needed its productions; we could not afford to retain animals; the earth had no longer need of them. Man found substitutes for the food and clothing he had formerly derived from the brutes; the space of land that would support an ox or a sheep could be made to produce more vegetable wealth in proportion to what could be gained from the animals.'[16] After that the construction platforms – enormous floating rafts – take up their positions at designated points across the oceans, the men obtaining their supplies of equipment and materials for the building of the sea-cities from fleets of aircraft. 'And so Man passed away from the land to the sea, using the former only to yield his requirements and the latter for residence, gliding about through the air in his aeromotives and looking back with pitying contempt upon the sordid grovellings of your age.'[17] The only possible attitude of the future world citizens, it seemed to Hay, would be wholehearted feelings of contempt for the failure of the Victorians to shape the world to their own desires.

These stupendous achievements were the victories of the applied sciences and of the united socialist states of the world. The rapid advances of the Century of Peace – that is, the twentieth century – had

> ... realised the dream of Socialist philosophers and Republican statesmen, and the countries of the Old World and the New, the races of civilised man, became absorbed into the United States of Humanity. The Confederate States of Australia, the Polynesian Republics, the Republic of South Africa, and all the countries where European or American influence was predominant, soon sent their delegates to the Grand Federal Congress – or Oecumenic Parliament as it soon came to be termed – and joined in the new political systems.[18]

So far all the hypothetical advantages of this ideal future world had been unquestioning and wholehearted acts of faith in the power of science and direct applications of potent contemporary ideas about the ideal society. And later on science decides one of the earliest actions of the United States of Humanity – a campaign of genocide on a scale that Hitler would have admired; for Hay, in his role as the Stalin of the people's paradise, was able to imagine the deliberate extermination of the coloured peoples of the world. His reasons were the purest social Darwinism: 'The duty of a rising race is either to absorb or to crush out of existence those with which it comes in contact, in order that the fittest and the best may eventually survive.'[19] It is scarcely necessary to say that this was not Darwin's opinion of human behaviour; it was, rather, what that wise and humane scientist feared. 'Looking to the world at no very distant date,' he wrote to William Graham in 1881, 'what an endless number of the lower races will have been eliminated by the higher civilised races throughout the world.'[20] And Hay realized that fear in the reasons he found for his imaginary acts of genocide:

> For it was seen to be impossible to raise the Chinaman to the level of the higher race; it was impossible to absorb him into it or fuse him with it. Even more obviously impossible was it to raise the Negro into a civilised and intellectual man, to make him fit for brotherhood with the Teuton and the Sclav. And Mankind was being welded under Social Government into one vast homogeneous uniform whole, and his increasing numbers were spreading abroad and filling up the spaces of the earth.[21]

For the security of United Humanity and for the sake of an abstract notion of equality the slaughter begins: the bombing planes pass over China, and 1,000 million human beings vanish in an instant of the imagination. But then, it seems that the Chinese were not truly human: 'We look back upon the Yellow Race with pitying contempt, for to us they can but seem mere anthropoid animals, not to be regarded as belonging to the race that is summed up and glorified in United Man.' The Negroes go the way of the Chinese and the Great Extermination is completed – the world has been made safe for the white nations in the name of science and for the noble cause of the brotherhood of man. 'What we chiefly learn from this page of history is the inevitable certainty of Nature's laws and the futility of resistance to them.' What response can there be to such deadly doctrines?

The forecasts in *Three Hundred Years Hence* at least have the value of demonstrating how these utopian myths seek to satisfy the most profound desires for peace and permanence; and the account of the Great Extermination serves to mark the extent of the changes that had taken place in the ideal states of the future since the appearance of Sebastien Mercier's *L'An 2440* in 1771. In the old tradition, which ran from St Thomas More to Sebastien Mercier, the laws of God and the intimations of Nature had decided the scheme of things in the just society; but, once the political theorists and the social scientists – Marx and Spencer, for example – had banished God from the universe, then the ideal states were quick to reflect the secular revolution in future societies that found the rational basis of their perfection in an unquestioning obedience to the laws of science and the promptings of intuition. The doctrinaire and authoritarian application of ideas in *Three Hundred Years Hence* was one of the first prophetic hints of the totalitarian despotisms that descended on Europe fifty years later. The brutal theories deployed in that projection of the 1880s and the real events in the history of the 1930s have shown that, if the whole of reality is considered to be concentrated in mankind and in human society, then every human being must conform to the requirements of the state. William Delisle Hay and Adolf Hitler had this much in common: the one wrote and the other acted in keeping with certain ideas selected from the science of their time; and they both saw their worlds in the light of most powerful myths about the destiny of a chosen race. In the archaic world, when men were the chattels of the gods, it was customary to ensure the peace of the state and the well-being of the people by the shedding of human blood. That prescription, written into the ancient mythologies, inaugurated the ideal state in *Three Hundred Years Hence* and almost succeeded in establishing *Das tausendjährige Reich* not so very long ago. In two pages of fiction Hay did to the Chinese and the Negroes what Hitler tried to do to the Jews. The written words anticipate the proclamations of the Nuremberg rallies: 'But those who were wont to extol the Mongols ... forgot the narrow mind incapable of lofty thought, the relentless cruelty, bestial vice, exaggerated animalism, and all the qualities that made the Chinese antihuman to the perceptions of the Elect of Nature.'[22]

What had been prefigured in the fiction of an ideal state was foreseen in the writings of Nietzsche. That demonic prophet had discovered the shape of coming things in the conditions of his time, especially in the spread of Darwinian ideas. He foresaw that, once Darwinism had wiped out the distinction between men and animals, an age of barbarism would begin. 'There will be wars such as have never happened on earth,' he wrote in 1888. In 1914 the European nations proved him right.

The engineering of a new world out of contemporary ideas

in *Three Hundred Years Hence* was only half the story in the tale of the future during the last three decades of the century. The other half came from the general experience of increasingly centralized government—the ministries, departments and boards brought into existence to take on the management of education and conscription, the oversight of public health, the supervision of mines and factories, the direction of the growing armies of state functionaries, of the new police forces, the new technical institutions, the new colonies overseas. A close and constant familiarity with these organizational solutions to the management problems of the industrial nations provided social models for the designers of the ideal states. As the utopists continued to find their political ideas in the theorists of their time, they began to give more and more attention to the administration of their world states and their advanced technological societies. The line of this development ran from the rudimentary proposals for world parliaments and continental unions in, for example, *In the Future* of 1875 and *The Dawn of the Twentieth Century* of 1882, to the detailed comprehensive scheme of the Samurai Administrators and world planning boards in Wells's *A Modern Utopia* of 1905.

This increasing concentration of interest on the problems and opportunities of the age was typical of utopian fiction during the last two decades of the nineteenth century. There was a falling off in the numbers of the self-congratulatory technological utopias, and an exceptionally rapid increase in the futuristic stories that set out to provide more practical and more political answers to the great social questions of the day. In the United Kingdom and in the United States the main burden of these stories was the reform and reorganization of industrial society; and this theme had the support of a growing body of ominous prophecies that warned against the dangers of despotic and ruthless governments in the future. Another development was the sudden explosion of utopias and dystopias in the United States that followed on the publication of Edward Bellamy's *Looking Backward* in 1888. It was the most widely read and most influential of all the nineteenth-century ideal states, because the lessons it proposed to teach the Americans had great relevance for the citizens of all the industrial nations. The roots of that famous story reach into the social conditions of the United States in the 1880s and into the temperament of the author. The rapid industrialization that followed on the Civil War, the consequent growth of population and of the great urban centres, the demands for better conditions of work, especially for legislation to stop the exploitation of child labour – these were the main factors that brought forth in that continent the new-style ideal state.

The first proposition came from Alfred Denton Cridge, who published *Utopia; or the History of an Extinct Planet* in 1884; and there he followed the workings of social justice from the industrialization of a planet to the subsequent social conflicts that disappear when the forces of reform abolish private enterprise and make equality in work and wealth the basis of the just society. The book appeared at the time when Bellamy was thinking over the many failings of the age and was discovering in Laurence Gronlund's *The Cooperative Commonwealth* many of the ideas for his own utopia. And then, towards the end of 1886, he sat down at his desk 'with the definite purpose', he said, 'of trying to reason out a method of economic organization by which the republic might guarantee the livelihood and material welfare of its citizens on a basis of equality corresponding to and supplanting their political equality.'[23] His guiding lights were his compassionate and democratic feelings, and these joined with his belief in a benign process of social evolution to give spirit and substance to his ideal state. As he said in a postscript, '*Looking Backward*, although in form a fanciful romance, is intended, in all seriousness, as a forecast, in accordance with the principles of evolution, of the next stage in the social and industrial development of humanity, especially in this country.'

So, in order to make his readers look forward, Bellamy brought the sleeper back to consciousness in the perfect socialist world of the year 2000. What that member of 'the pampered Bostonian aristocracy', Julian West, discovers is that political theory has transformed the free-enterprise system of the technological utopias into an egalitarian world state. West reports that

> ... the great nations of Europe as well as Australia, Mexico, and parts of South America, are now organized industrially

> like the United States, which was the pioneer of the evolution. The peaceful relations of these nations are assured by a loose form of federal union of world-wide extent. An international council regulates the mutual intercourse and commerce of the members of the union and their joint policy towards the more backward races, which are gradually being educated up to civilized institutions.[24]

The index to the British edition (*No 1 – The Bellamy Library*) lists the principal facts in this ideal existence: Advertising, Lack of; Army and Navy abolished; Army Industrial, its perfect organization and self-devotion; Art, how encouraged; Buying and Selling, anti-social; Brotherhood of Man, belief in; Capital, growing concentration of; Children, maintained by the State; Churches, voluntary; Christianity, nominal only; Competition and Co-operation compared. And so on through all the elements of the perfect socialist world to the final entry: Women, freedom from household cares; form part of the industrial army.

Within a year of publication *Looking Backward* had gained a degree of influence and reputation everywhere that matched Chesney's triumph with the *Battle of Dorking*. It became a sacred text in the holy war against capitalism; and it disproved Sorel's contemptuous remarks about the ineptitude of utopian literature, since the future that Bellamy framed in his prophecy had a major influence on the socialist movement in the United Kingdom. The book sold in hundreds of thousands; there were imitations and counterblasts throughout Europe and America; clubs were formed to propagate Bellamy's message and within three years 163 of them had spread across the United States. *Looking Backward* was translated into most of the European languages – even Russian; and in 1891 the Austrian scholar Friedrich Kleinwächter reported that 'today it is read in almost every village'. In England an article in the *Contemporary Review* for January 1890 discussed the significance of the quite exceptional interest in a poorly written narrative that for long passages read like a Blue Book on industrial reforms:

> The rapid and extraordinary success in all the Anglo-Saxon world of Mr Bellamy's book – 240,000 copies sold in the

> States, and 40,000 in England at this date ... is a symptom well worthy of attention. It proves that the optimism of old-fashioned economists has entirely lost the authority it formerly possessed. It is no longer believed that, in the virtue of the 'laissez-faire' principle, everything will arrange itself for the best in the best of all possible worlds.[25]

The secret of Bellamy's success was that he had invented the working model of an alternative industrial society; he faced all the great social issues of the day, and he showed how his solutions met the necessary conditions of prosperity and progress in a just society that was an America changed for the better, but not changed out of all recognition. Bellamy had composed his utopia at the moment when the great industrial nations were moving towards the second phase of social reform – old age pensions, national insurance schemes, higher education – that began to lay the foundations of the modern world in the early twentieth century. How central his thinking was to reforming opinion in the United Kingdom can be seen in the proposals for national renewal that the Liberal politician C. F. C. Masterman made at the turn of the century. Masterman believed that 'a menace to the future progress of humanity' threatened the nation, and he found the greatest danger in 'the old astonishing creed that if each man assiduously minds his own business and pursues his own individual advancement and the welfare of his family, somehow by some divinely ordered interconnections and adjustments the success and progress of the whole body politic will be assured.'[26]

Bellamy's ideal commonwealth was one of three classic solutions to the burning questions of social justice and urban organization. The other two appeared with appropriate promptitude in 1890 – one from an English poet and social reformer, the other from an Austrian editor and economist. In the eloquent and often moving descriptions of the good life in the *News from Nowhere* William Morris suggested, rather than proposed, that all the most difficult problems would be solved if society moved backward from industrialism to the co-operative communities of a socialist arcadia. Theodor Hertzka set out the contrary doctrine in *Freiland, ein soziales Zukunftsbild*, where he demonstrated that the advance of capitalism would lead to a better life for all. Hertzka was a distinguished Viennese economist of the laissez-faire school; and before he became internationally famous after the publication of *Freiland*, he had already gained a national reputation as economics editor of the *Neue Freie Presse* and as the founder and first editor of the *Wiener Allgemeine Zeitung*. His proposals came straight from the interests of an intelligent capitalism—let a state bank provide the resources for private enterprise not in cash but in the form of plant and installations, and let the state recover the outlay from the successful entrepreneurs so that the process of loans and profits will promote the prosperity of all. Hertzka placed his experimental enterprise in Kenya, and in his story volunteers by the thousand set off to prove Hertzka's points and to make their fortunes in building up a prosperous economy. The effects of Hertzka's scheme were, once again, beyond all expectations; and the capitalism of *Freiland* attracted as much interest as the socialism of *Looking Backward*. Indeed, Hertzka enjoyed a greater popularity in Europe than Edward Bellamy: his book went into many foreign translations and many editions; committees sprang up to promote his ideas, and a party of devoted followers actually set out to find an area in Kenya where they hoped to establish a truly profitable society.

The profit-and-loss motive and the careful disposing of human beings in *Freiland* were anathema to William Morris, who had spent a most industrious and successful life in trying in his art and poetry to reflect the glories of an idealized medievalism on the darkness of industrial England. Morris looked for the harmony and the happiness he felt to be natural to human existence; and he concluded that, since the industrial revolution had degraded and dehumanized the quality of life, the only hope for the future was to change society itself. Socialism was, therefore, the road to the earthly paradise; and in a lecture at the shrine of laissez-faire economics, Manchester, he explained the political principles that had led him to socialism: 'So, then, my ideal is first unconstrained life, and next simple and natural life. First, you must be free; and next you must learn to take pleasure in all the details of life: which, indeed, will be necessary for you, because, since others will be free, you will have to do your own work. That is in direct opposition to civilization, which says: Avoid trouble, which you can only do by making other people live your life for you.'[27] These views touched off another great explosion in the history of utopian fiction. As soon as Morris had read *Looking Backward*, he dismissed the book in an unfortunate phrase as 'a Cockney paradise' and then denounced it in a review:

> In short, a machine-life is the best which Mr. Bellamy can imagine for us on all sides; it is not to be wondered at then that his only idea of making labour tolerable is to decrease the amount of it by means of fresh and ever fresh developments of machinery. Mr. Bellamy worries himself unnecessarily in seeking (with obvious failure) some incentive to labour to replace the fear of starvation, which is at present our only one, whereas it cannot be too often repeated that the true incentive to useful and happy labour is and must be pleasure in the work itself.[28]

Work and pleasure provide the characteristic images in the *News from Nowhere*, since Morris gives most of his attention to shaping episodes that reveal the quality of the better life. The book is a dream, an epoch of rest as the subtitle says; and the dreamer of dreams sets the crooked straight, imagining the kind of paradise that would obliterate the hell of industrialism. Like Jefferies and Hudson he wipes out contemporary England. His dreamer falls asleep on a winter's night and awakens on a perfect summer morning to discover that a great transformation had taken place: 'The soap-works with their smoke-vomiting chimneys were gone; the engineer's works were gone; the lead works gone; and no sound of rivetting and hammering came down the west wind from Thorneycroft's.' The filthy Thames runs clean again, and the hideous England of 1890 has given place to a land of green fields, beautiful houses, handsomely dressed people, and cottage crafts. Machinery has apparently vanished—although there is a passing reference to 'force vehicles', most of the work seems to be done by hand in Banded Workshops. The deliberate arcadianism never fails to bring out the slide-rules of the critics; and even Morris's one-time disciple, Graham Wallas, missed the point when he reached for his notebook and 'made a rough calculation that the citizens of his commonwealth, in order to produce by the methods he advocated the quantity of beautiful and delicious things which they were to enjoy, would have to work two hundred hours a week.'[29] The point of the parable appears in the last paragraph, when the people of the future send the sleeper back to the London of 1890: 'Go back, then, and while you live you will see all round you people engaged in making others live lives which are not their own, while they themselves care nothing for their own real lives – men who hate life though they fear death.'

The unprecedented outpouring of utopian prophecies in the last decade of the nineteenth century reinforces the point made earlier in this chapter: that the tale of the future came to the first full perfection of its form in the 1870s, when it was accepted as the one mode of communication that could deal in a popular and effective way with the ambitions and anxieties peculiar to an era of ceaseless change. By the 1890s the tale of the future had reached a position of world-wide influence: Jules Verne in translation for the Japanese and Chinese; Bellamy read in many editions from Brisbane to British Columbia; tales of future wars in Arabic and Turkish. These were the first signs of a growing realization throughout the planet Earth that modern technology was bringing all nations ever closer together, that the evolution of industrial civilization called for a new order of human society, that future wars could affect all the peoples of the world. Moreover, the appearance of the first popular essays in futurology in the 1890s was a sign that urban mankind was beginning to realize it had passed through the old time zone of then-and-now and, therefore, had to prepare for the many changes to be expected in tomorrow's world. And in the mysterious way of human history in the decade when the new practice of prediction had begun to take over some of the operations of futuristic fiction, the idea of the future found its high priest in H. G. Wells, who did more than any other writer to establish the awareness of the future that is so characteristic of the twentieth century.

Between 1895 and 1905 – that is, during the last ten years of Verne's life – Wells had earned a world reputation as being a more inventive, more gifted, more varied and far more influential writer than Verne had ever been. By 1905, the year of Verne's death and forty years after the publication of Verne's first futuristic story, *From the Earth to the Moon*, Wells took to legislating for the future of mankind in *A Modern Utopia*. In his perceptive and analytical way he began with the problem that faced all mankind in the first decade of the twentieth century:

> The almost cataclysmic development of new machinery, the discovery of new materials, and the appearance of new social possibilities through the organized pursuit of material science, have given enormous and unprecedented facilities to the spirit of innovation. The old local order has been broken up or is now being broken up all over the world, and everywhere societies deliquesce, everywhere men are afloat amidst the wreckage of their flooded conventions, and still tremendously unaware of the thing that has happened.[30]

Wells went on to outline the plans and aspirations that would bring about the world state and world peace. 'It would be so easy to bring about a world peace within a few decades,' he wrote. All that was needed was 'the will for it among men. The great empires that exist need but a little speech and frankness one with another.'

7

Possibilities, Probabilities and Predictions

The preceding chapters have traced the different ways in which the idea of the future found specific means of expression in the scientific romance, in the tale of the war-to-come, and in the utopian prophecy of the ideal state. These distinct modes of writing about the future were specialized responses to the changing conditions of life in the great urban societies. They had their roots in the interests of social groups: the science fiction story was the creation of professional authors; the colonels and the military correspondents trumpeted their hopes and fears through the tale of imaginary warfare; and the designing of perfect future worlds was the concern of all who knew what was good for society.

As these tales of the future advanced along their separate evolutionary paths, there was a comparable process of development in the art of forecasting the changes to be expected in the years ahead; and, as this chapter will seek to show, the new practice of prediction was the creation of yet another group of professional writers. These were the first social workers and town-planners, the statisticians, civil servants, the sociologists and scientists, and all those earnest citizens whose education and experience gave them a vocational interest in the social or technological potentialities of an industrial civilization. Their chosen form of writing was not fiction but the predictive essay; and they addressed themselves to the readers of the professional journals and the middle-class magazines. From the beginnings of the new literature about 1870 these forecasts proved a more objective, more rational and often more comprehensive method of composing the pattern of coming things than was possible – or even desirable – in the elementary adventures of the science fiction stories, or the self-regulating prophecies of the imaginary wars, or the self-evident demonstrations of the ideal states. Because these forecasts were more demanding and far more complex than the contemporary tales of the future, they were slow to discover their most suitable means of communication and they were the last to attract the general interest.

In origin they go back to the first half of the nineteenth century, when the earliest histories of the new technologies introduced descriptions of the next stage to be expected in the onward movement of the industrial nations. These incidental forecasts were limited in their scope and rudimentary in their methods. Like the contemporary tales of the future, they had little of importance to say until the 1860s, when the new ideas about the evolution of society encouraged a growing interest in the many factors that made for future change. Thus, the 1860s saw the beginnings of modern science fiction in the predictive epics of Jules Verne; and during that decade the earliest technological utopias appeared, the first confident attempts to divine the most likely state of urban society in the coming centuries.

This association between ideal theories and optimistic expectations remained close and constant throughout the tale of the future and the literature of prediction up to the outbreak of the First World War. Although some writers rejected the dominant belief in the manifold advantages of material progress – W. H. Hudson and William Morris, for example – the majority held to the perfectionist conviction that society was developing according to an evolutionary and adaptive code. Because this belief was more persuasive and far more pervasive than the original and simpler belief in technological progress, it had become common practice by the 1860s for writers to project the various dogmas of their time – social or political, military or imperial – in anticipations of things-to-come. For instance, on the occasion of the Paris Exhibition in 1867 Victor Hugo accepted an invitation to write an article about the history and future of Paris for the *Paris Guide*. In his introductory remarks on 'The Future' he renewed for the visitors to the exhibition the promises of Saint-Simon and Lamartine in his forecast of the glorious epoch of European peace and unity: 'In the Twentieth Century there will be an extraordinary nation. This nation will be illustrious, wealthy, thoughtful, pacific, cordial to the rest of mankind ... The capital of this nation will be Paris, but its country will not be known as France. In the Twentieth Century its country will be called Europe, and in after centuries, as it still and ever develops, it will be called Mankind.'[1]

Whilst the Parisians read their Victor Hugo, an Oxford audience heard another poet talk about the future of mankind. In the May of 1867 Matthew Arnold ended his ten-year term as Professor of Poetry with an attack on the enemies of culture, and he continued the battle for sweetness and light in five consecutive articles in the *Cornhill Magazine* in 1868. Victorian society was engaged in a titanic struggle between culture and anarchy, Arnold believed, and the issue would decide whether that society would ever subdue the forces of privilege, prejudice and ignorance. Arnold urged his readers to keep on 'trying to find in the intelligible law of things a firmer and sounder basis for future practice'. The objective of all citizens should be to encourage the free play of the mind, for that alone could bring about the spiritual transformation of the Barbarians, Philistines, and Populace. And so, 'believing in right reason, and having faith in the progress of humanity towards perfection, and ever labouring for this end, we grow to have clearer sight of the ideas of right reason, and of the elements and helps of perfection, and come gradually to fill the framework of the State with them.'[2]

One year after Arnold had outlined these hopes for the future of society, Sir John Seeley became Regius Professor of Modern History in the University of Cambridge, and from his first lectures he made it his business to impress on his young gentlemen the value of history as an aid to prediction. He argued that the study of history 'should not merely gratify the reader's curiosity about the past, but modify his view of the present and his forecast of the future ... it ought to exhibit the general tendency of English affairs in such a way as to set us thinking about the future and divining the destiny which is reserved for us.'[3] The professor and the poets had drawn on the European ideology of progress to express the most powerful beliefs of an international urban society then in the middle phase of a world-wide expansion. In their various ways they bore witness to the universal effects of an infallible evolutionary system. Like Herbert Spencer they found an eternal link between past causes and future effects; and they could affirm with Herbert Spencer that, 'after observing how the processes that have brought things to their present stage are still going on, not with a decreasing rapidity indicating approach to cessation, but with an increasing rapidity that implies long continuance and immense transformations; there follows the conviction that the remote future has in store, forms of social life higher than any we have imagined.'[4]

A marked element of social thinking, a desire to foresee the evolution of the nation or of the world, a still unshaken faith in material progress, and a general eagerness to communicate expert opinion – these were the principal characteristics of the forecasts that appeared in Europe and North America between 1870 and the outbreak of the First World War. It was a time when the industrialized countries turned to considering the state of the nations and the direction in which the world seemed to be moving after the war between the French and Germans, after the war between the States, after the British had begun to comprehend the scale of American and German industrial expansion. New universities and technical colleges, new professions and many new journals were the first and most evident results of this search for means of meeting the demands of new social conditions. The foundation of the Massachusetts Institute of Technology in 1865 and of the Institution of Electrical Engineers in 1871 were unmistakable signs of the increasingly intimate ties between industry and education, between society and the technological professions. And in 1862 a grateful nation consecrated the immense utility of this relationship with the new word of *specialist*: 'One who specially or exclusively studies one subject or one particular branch of a subject.'

In the field of engineering, for example, the rapid growth of specializations – another new word in 1865 – had led to the founding of new associations dedicated to applying professional knowledge within distinct areas of technology – chemical, municipal, sanitary, marine, mining, water, heating and ventilation. As the professions multiplied in keeping with the division of intellectual labour, there was a corresponding growth in the accounts of the changes to be expected in the next decade or the next century. These soon made an increasingly frequent appearance in the middle-class periodicals from 1870 onwards; for the practice of prediction followed out of a dialogue that had developed – and still continues to develop – between the expert writer and the literate citizen. It was a considered and professional response to major questions about the shape of coming things; it was a spontaneous development from within Victorian society that has many parallels with the explosion in social and technological forecasting during the 1960s.

The extent of this advance from fiction to prediction can best be assessed by looking into the changes that took place in the two very different fields of the town planning movement and the forecasting of future wars. First, then, it is a matter of historical fact that the constant increase in population required the making of demographic forecasts, and that the consequent expansion of urban communities was a positive encouragement for radical proposals to reorganize the industrial cities. In 1870 the director of the Bureau of Statistics in Paris, Alfred Legoyt, made a start on this problem in his *Du Progrès des agglomérations urbaines*. By 1898 this methodical investigation of town and country problems had produced an even more exhaustive analysis in the statistical work of Paul Meuriot, *Des Agglomérations urbaines dans l'Europe contemporaine*. Meuriot showed that the railway and the steam-engine had brought about the evils of rural depopulation and urban sprawl, and by those same tokens the new technology of the dynamo and electric transport would solve the more serious problems of the vast industrial cities:

> Already the ease of transportation, a thing unknown to our predecessors, makes it possible to reduce the excessive population of the towns; and another innovation, the transmission

> of electrical power, can have quite different results. The first would be to keep the industrial worker in his own home and to do away with those concentrations of men that the modern factory system makes necessary ... In a word, one scientific revolution caused the extreme development of the urban community, and in the same way another scientific revolution will produce the remedy for these evils of excessive growth.[5]

The advance from theory to practice followed within a year; and in 1899 the then unknown student of the École des Beaux-Arts in Paris, Tony Garnier, began his life-long labours to transform the statistics of urban development into the architectonics of his rational, socialist and utopian design for *La Cité industrielle* – a planned harmony of administrative and educational sections, sports centres, residential areas, all separated by green belts and set in inter-related zones of parkland, with the factories and the railway station located away from the rest of the city.

The same problem of urban expansion produced similar solutions in the United Kingdom. During the meeting of the British Social Science Association in the October of 1875 the President of the Health Section, Benjamin Ward Richardson, gave an address which proved to be most influential after it was published as a pamphlet in 1876 with the title of *Hygeia – a City of Health*. Richardson began from the fact of a common professionalism: 'We meet in this Assembly, a voluntary Parliament of men and women, to study together and to exchange knowledge and thought on works of everyday life and usefulness.' In the exalted language of a new, secular priesthood the orator went on to assert that professional men and women had a special responsibility towards posterity. Their expertise had set them apart from the masses of the nations, and with their profound knowledge of the causes at work in society there came the gift of foresight: 'We are privileged more than any who have as yet lived on this planet in being able to foresee, and in some measure estimate, the results of our wealth of labour as it may be possibly extended over and through the unborn ... we are the waves of the ocean of life, communicating motion to the expanse before us, and leaving the history we have made on the shore behind.'[6] Richardson considered that sanitary engineering and medical science had dictated the best conditions for city life – a population of 100,000 living in 20,000 houses on 4,000 acres of land – and he predicted that careful attention to his planning arrangements would raise the levels of life expectancy and of general health in the future city. His words compose one of the early acts of faith in social prediction and town planning.

> These thoughts on the future, rather than on the passing influence of our congressional work, have led me to the simple design of the address which, as President of this Section, I venture to submit to you today. It is my object to put forward a theoretical outline of a community so circumstanced and so maintained by the exercise of its own free will, guided by scientific knowledge, that in it the perfection of sanitary results will be approached, if not actually realised, in the co-existence of the lowest possible general mortality with the highest possible longevity. I shall try to show a working community in which death – if I may apply so common and expressive a phrase on so solemn a subject – is kept as nearly as possible in its proper or natural place in the scheme of life.[7]

Behind the flatulent prose an acute and original mind was at work; for Richardson was one of the first generation of professional writers who found in contemporary science the blueprints for a better kind of life in the future. His career is an early case-history in the movement for the managing of social and technological change. In 1847 he entered the medical school of Anderson's University in Glasgow, where he would undoubtedly have experienced the decided social and practical interests of the only major technological institution at that time in Great Britain. Ten years later Richardson was one of the first lecturers in the new field of public hygiene, and ten years after that he was elected a Fellow of the Royal Society.

When this eminent Victorian appeared before the Social

Science Association in 1875, he pursued a line of argument that was common to the many propositions – about the future of war, the navy, education, religion, the colonies – then appearing in pamphlets and periodicals. Richardson went from his analysis of the conditions of life in the industrial city to an anticipation of the benefits that would follow from his scheme for segregated traffic, regular open spaces and strict building controls. His plan became a primary text in predictive literature and it proved to be an influential essay in the theory and practice of town-planning. Jules Verne, for example, was quick to transform the theories of *Hygeia* into the model city of Franceville in *The Begum's Fortune* of 1878; and there were similar borrowings in Bellamy's description of the ideal Boston of *Looking Backward*: 'Miles of broad streets, shaded by trees and lined with fine buildings, for the most part not in continuous blocks but set in larger or smaller enclosures, stretched in every direction. Every quarter contained large open squares filled with trees, among which statues glistened and fountains flashed in the late afternoon sun.'

This chain of original ideas and specific applications was typical of the plans and predictions that looked to the future of industrial society. As the utopian fiction of Verne and Bellamy showed, there was an international traffic in ideas about the best plan for the city and for human society – a fact that is apparent in the most influential of all the foundation documents in the history of the town-planning movement, *To-morrow: A Peaceful Path to Real Reform*, which Ebenezer Howard published in 1898. From *Looking Backward* Howard drew what were for him important conclusions about the moral and practical basis of the just society, and in *Hygeia* he found a useful model for presenting his project of the planned community which would represent a balance between homes and industries, between individual enterprise and municipal control. Howard opened with the menace of vast cities and rural depopulation, and he closed his introduction with a famous passage on 'The Three Magnets', or the advantages and the disadvantages of life in the town and in the country. The argument that runs through everything is the need to understand the unique conditions of a technological civilization:

> So marked are the changes which society exhibits – especially a society in a progressive state – that the outward and visible forms which our society presents today, its public and private buildings, its means of communication, the appliances with which it works, its machinery, its docks, its artificial harbours, its instruments of war and its instruments of peace, have most of them undergone a complete change, and many of them several complete changes, within the last sixty years.[8]

Howard had looked the sphinx in the eye, and he knew that the beginning of the solution to the many problems of the urban environment was to accept the realities of the factory and to acknowledge the human needs of the citizens. He traced his path towards the practicable future of 'the social cities' by reasoned stages – revenue, expenditure, administration, construction – and the continuing argument of his wise and rational projection was to reiterate the priorities of order, proportion and the human scale in all things. The crowded cities of the Victorian expansion were no longer tolerable; for 'they were the best which a society largely based on selfishness and rapacity could construct, but they are in the nature of things entirely unadapted for a society in which the social side of our nature is demanding a larger share of recognition.' His solution was to reduce the size of the great cities, especially of London, and to build towns in which the populations would be limited and the location of the homes, shops, schools, parks, factories and railways could be carefully planned. So, Ebenezer Howard called on the British people to bend themselves 'to the task of building up clusters of beautiful home-towns, each zoned by gardens, for those who now dwell in crowded, slum-infested cities. We have seen how *one* such town could be built; let us now see how the true path of reform, once discovered, will, if resolutely followed, lead society on to a far higher destiny than it has ever yet ventured to hope for, though such a future has often been foretold by daring spirits.'[9]

The prophet found disciples throughout the world. In 1899 Howard established the Garden City Association and in 1903 the founding fathers started on the planning of the first garden city of Letchworth. That small beginning was a specific English contribution to the growing body of future-directed thinking that aimed to reorganize the industrial city. In 1899 the benign anarchist and expert geographer Prince Kropotkin had already put forward his scheme for smaller industrial units and the planned development of rural areas in a famous book on *Fields, Factories, and Workshops: or, Industry combined with Agriculture, and Brainwork with Manual Work*. Again, in 1905 a Minister of Agriculture and one-time Prime Minister of France, Jules Méline, published his plan for the improvement of the nation, *Le Salut par la terre*. He pointed to Bourneville and Port Sunlight as admirable models of the organic community and presented his case for a large-scale movement from the towns back to the land. Then came the Town Planning Conference of 1910, when for the first time British architects met to consider professional remedies for the many failings of their predecessors. In 1919 Howard founded his second garden city at Welwyn, twenty-three miles north of London, and lived there until his death in 1928. He was an early exemplar of the active, intelligent and concerned citizen, a David in the Goliath world of powerful industrial organizations and the environmental disaster areas of the Black Country, the Ruhr, and Detroit. Had he lived on until the Town and Country Planning Act of 1944, Sir Ebenezer Howard would have known that the prophet can be heard as well as honoured in his own country.

Howard laboured all his days to establish the principle of a planned environment for industry and individual lives, but not alone. He was part of a commonwealth of original minds that looked to the future to resolve the many dilemmas of the late Victorian world. It is, therefore, a mistake to think of separating Howard and his scheme for social cities from the prophetic utopias of Bellamy, Hertzka, and even of William Morris; and it is equally wrong to set his insights apart from the very different considerations that prompted the outpouring of stories about future wars. The visionary capitalism in Theodor Hertzka's *Freiland* and the socialist harmony of Edward Bellamy's *Looking Backward* had this much in common with the proposals of Howard and of all the others – even the generals and the political propagandists – who called on their contemporaries to prepare for the changes that would have to come: they were all most keenly aware of the universal flux in which all things moved and many of them were eager to play their part in directing society towards new horizons in the future. From the 1860s onwards the whole bent of writing about the future, in fiction and in prediction, was a trying out of modes for presenting ideas about coming things and a search for devices to bring about those radical changes – in politics, the army, urban life, or in Ireland – that would rescue the twentieth century from the potential dangers and real calamities of the nineteenth. All these prophecies and predictions had their sources in the collective consciousness of contemporary society – in the central experience of universal progress that caused some to predict the technological advances still to come and others to forecast the necessary changes of direction in the forward movement of mankind.

Despite the many fundamental differences in idea and method that ran through the prophecies and the predictions, there were at the same time marked similarities; for they all came white-hot from a world shaken by the volcanic eruptions of changing conditions. For instance, there were those affinities between the anticipatory adventures in the science fiction of Jules Verne and Chesney's classic account of future warfare in the *Battle of Dorking* that have already been discussed. Again, a comparable relationship made Jules Verne the literary half-brother of Ebenezer Howard. Both men owed their success to the new means of communication, Verne to the invention of the illustrated magazine for adolescents and Howard to the mass press and later to wireless and the cinema. In like manner, Verne found a career in creating vivid projections of future achievements and Howard devoted his life to a project for the reorganization of the industrial city.

Verne and Howard, Chesney and Bulwer-Lytton, Hertzka and Bellamy – the work of these writers represented the range of territorial interests in the new-found land of the future. Their descriptions and prescriptions marked the furthest points of advance in the exploration of the changing relationship between man and nature, between the individual and his society. Imaginary wars, scientific romances, utopias, dystopias, plans and predictions – whatever the differences in intention may have been, the authors were all agreed that future changes would have immense effect throughout the industrialized societies of the world. For Jules Verne the future was a licence to anticipate the limitless opportunities of an obedient technology; for Howard it brought a serious call to personal adjustment and national adaptability. 'But are the people of England to suffer for ever', Howard asked, 'for want of foresight of those who little dreamed of the future development of the railways? Surely not. It was in the nature of things little likely that the first network of railways ever constructed should conform to true principles; but now, seeing the enormous progress which has been made in the means of rapid communication, it is high time that we availed ourselves more fully of those means and built our cities upon some such plan as that I have crudely shown.'[10] The corollaries of Howard's thesis were that 'each generation should build to suit its own needs', and that the reader should not 'take it for granted that the large cities in which he may perhaps take a pardonable pride are necessarily, in their present form, any more permanent than the stagecoach system.' Howard's conclusions demanded a total change of mind: 'Can better results be obtained by starting on a bold plan on comparatively virgin soil than by attempting to adapt our old cities to our newer and higher needs? Thus fairly faced, the question can only be answered in one way; and when that simple fact is well grasped, the social revolution will speedily commence.'[11]

There had been many revolutions – in education, in technology, methods of warfare, means of transportation, political theories, the conditions of city life – and these in one way or another were always the points of origin and the principal considerations in the unprecedented upsurge of forecasts during the last three decades of the nineteenth century. In the continuing debate about new towns, new wars, new machines and new societies the guiding slogans were always the same – evolution and adaptation. These ideas were, for example, central both to the description of future warfare in the *Battle of Dorking* and to the forecast of 'War in the Twentieth Century' in *Anticipations* by H. G. Wells; they were the products of a period of catalytic action, when various factors combined to

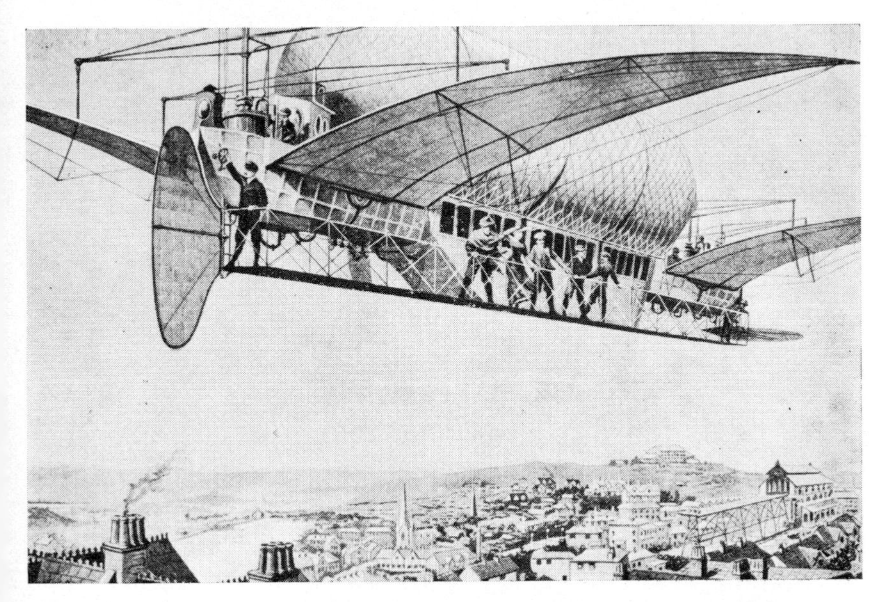

26 For many the most desired achievement of the future would be rapid aerial flight.

27 And for many others it seemed natural to apply scientific advances to the conduct of war.

28–31 The vision of the future meant the possibility of planetary communication, of the fighting vehicles of Wells's Martian invaders, or gold-mining on the Moon, even a world covered by the Red Weed from Mars.

32 The end of the world, or the end of civilization, was a possibility that attracted the interest of some writers.

33–34 Illustrated magazines entertained their readers with projections of coming things—sun radiators and the delights of life in housetop gardens.

35–36 One anticipation was the possibility of a harsh autocratic government based on highly developed technologies that would control all workers in a rigorous economic scheme.

37 When imagination was pushed to the limit of the possible, one result was the entirely logical fantasy of the Selenite civilization in Wells's *The First Men in the Moon*, 1901.

38–39 The rapid growth of Victorian cities prompted visions of the future metropolis and of great cities of the sea floating in mid-Atlantic.

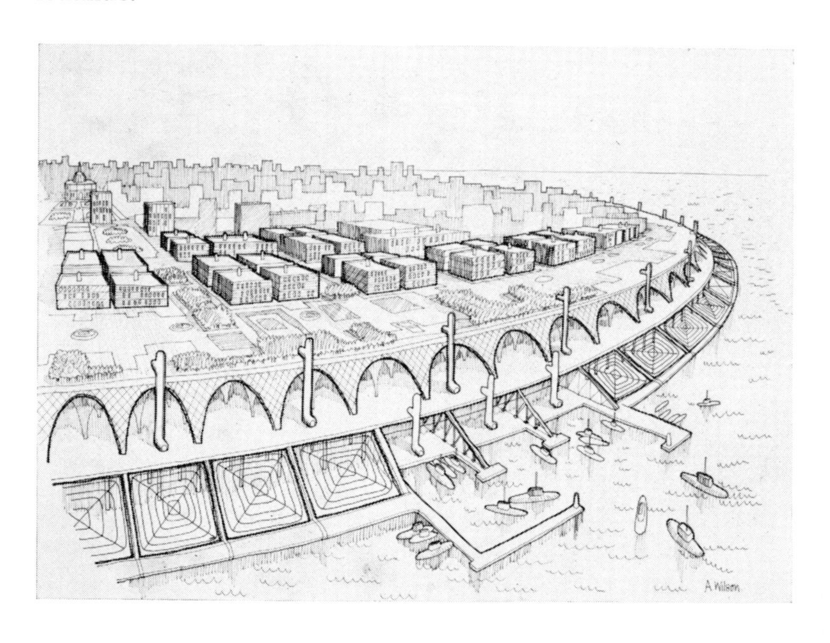

40 A bridge across the Channel, talked of for half a century, becomes reality in an artist's impression.

41–42 The great engineering enterprises of the future could be the fantasy of a popular writer, and they were often the calculations of practising engineers.

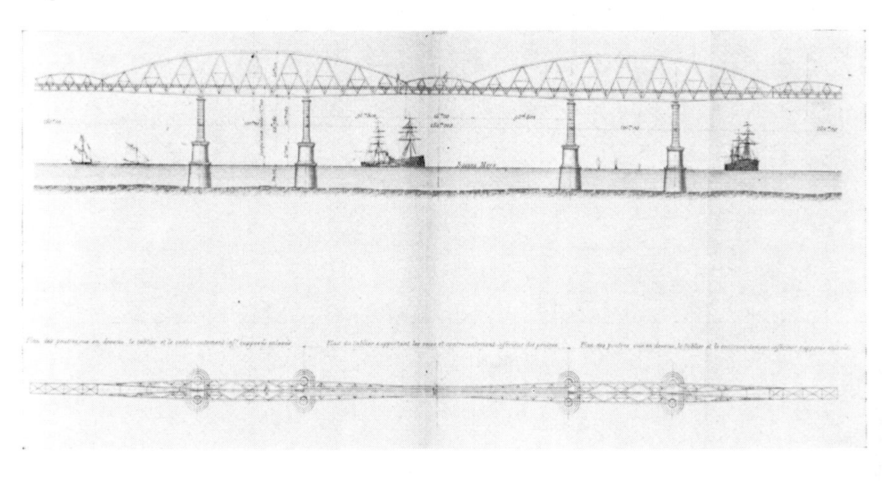

promote the discussion about the need to prepare for, or to guard against, the changes that were sure to come. After 1870 the more important of these influences in the rapid expansion of the new predictive literature were: the recent diffusion of sociological theories, the increasing application of statistical methods to the investigation of urban conditions, and the constant advances in literacy and in education. These developments had the double effect of encouraging a demand for more expert information about possible changes and of obliging the propagandists to appeal directly to the electorate.

The principal periodicals responded to this interest in the future by providing an increasing number of articles in which statisticians and colonels joined with sociologists and politicians in forecasting the shape of things-to-come for the readers of *Blackwood's Magazine*, the *Contemporary*, the *Cornhill*, the *Fortnightly Review*, *Macmillan's Magazine*, *Fraser's Magazine*, and many other middle-class journals. They wrote with the knowledge and assurance of experts: W. T. Stead on 'The Future of Journalism', Canon Kingsley on 'The Natural Theology of the Future', Chesney on 'The Military Future of Germany', Francis Galton on 'Stanley's Discoveries and the Future of Africa', Camille Flammarion on 'The Last Days of the Earth', Moritz Kaufmann on 'The Society of the Future', Herbert Spencer on 'The Coming Slavery'.

One of the more interesting examples of this development in the communication of specialist knowledge came from Alphonse de Candolle, the scientist son of the celebrated botanist. In 1870 he produced a statistical inquiry into the 'Advantage for Science of a dominant language and which of the modern languages will necessarily be dominant in the Twentieth Century.' He started from the proposition that, because there had been so many major advances in science in Northern Europe, any dominant language of the future would have of necessity two characteristics: '(1) It would have to contain sufficient Germanic and Latin words and structures to be available both to the Germans and to those who speak one of the Latin languages; (2) It would have to be spoken by a majority of the civilized world.' The patterns of the English language, de Candolle claimed, were '... adapted to modern tendencies. Whether it is necessary to steer a ship, stop a train, strip down a machine, carry out an experiment, or speak briefly to hard-working and practical people, it is the one outstanding language.' Given the projected increases in national populations, it followed that by 1970 English would be the dominant language of industrial society throughout the planet:

> Thus, the three principal languages spoken today will in a century have made the following advances:
>
> The English language will have advanced from 77 to 860 millions.
>
> The German language ,, ,, ,, ,, 62 to 124 millions.
>
> The French language ,, ,, ,, ,, 40½ to 69½ millions.
>
> Those speaking German will make up a seventh and those speaking French a twelfth or thirteenth part of the English-speakers; and these two languages together will not be one *quarter* of those who speak English! The French and German languages will then have the same relationship to English as Dutch and Swedish now have to them.[12]

Since the use of any language is related to the increase in population, the future dominance of the mother-tongue of the American, Australian and British peoples was assured. The moral in all this was that the English-speaking nations had a responsibility to the rest of the world: 'It is their duty, and it is in their interest, to preserve the actual unity of the English language, whilst allowing for all necessary or desirable modifications that will be made by common agreement.' And then there was Oscar Wilde who said that the British and the Americans had everything in common except language.

This new gospel of good works and preventive actions was a guiding light of predictive literature. Even the most casual examination of the links between contemporary tendencies and anticipated developments usually led to conclusions about the responsibility of the citizen or of the nation for the proper management of future change. For instance, the habit of sounding a general alert for posterity was a necessary device in the literature of imaginary warfare both in Europe and in North America. This unique mode of communication began as an immediate British reaction to the formidable demonstration of German power in the war of 1870, and within one decade it had become an important means of conveying messages from the captains and colonels to the citizens and taxpayers of the industrial nations. The closed-circuit style of their narratives (forewarned is forearmed) and the close relationship between author and public (pay or perish) led to an unusually self-contained and purposeful form of writing; and this sometimes had extraordinary effects in the way, for example, the *Battle of Dorking* or *La Guerre anglo-franco-russe* could work upon the hopes and fears of nations.

These descriptions of the war-to-come were labours of love, offerings to the nations from distinguished professional men who wished to present their expert knowledge and earnest recommendations in the most effective manner possible. So, Sir George Tomkyns Chesney, founder of the Indian Civil Engineering College, wrote the notorious *Battle of Dorking* to concentrate national attention on the case for conscription; Sir William Butler, the protégé of Lord Wolseley and the friend of Victor Hugo, wrote *The Invasion of England* as a serious warning against the dangers of national degeneracy; Hugh Arnold-Forster, the favourite nephew of Matthew Arnold and a Secretary of State for War, wrote *In a Conning Tower* 'to illustrate the results of a long course of careful observations of modern naval progress.' The book of records ended with the many productions of Capitaine Émile Driant, son-in-law of General Boulanger and later deputy for Nancy, who turned out seven stories of 7,616 pages in all in a series that went from *La Guerre de demain* in 1889 to *La Guerre souterraine* in 1913.

Because these accounts of possible wars formed the most coherent body of writing in futuristic literature up to 1914, they provide the clearest illustration of the working of those special conditions that brought about the consolidation of social and technological forecasting in the 1890s. First, the range of the discussion of future wars followed a widening curve of distribution; it started with specialist fiction from serving officers but in the 1890s it included forecasts from the growing corps of military and naval correspondents who wrote about The Future of Maritime Warfare, The German Peril and The Future of the Northwest Frontier. Second, as the field of communication widened, there was a corresponding increase in the input of new ideas; and out of this recognition of the many interconnexions in industrial society the practice developed for writers to look beyond immediate military considerations to the social or political consequences of any future war. Third, the intention in this communicating of expert knowledge was to persuade the public of the need for radical changes. Colonel Maurice who collaborated with Admiral Colomb in writing *The Great War of 189–*, was completely candid on this point: 'The audience to which I am anxious to appeal are the lawyers, the business men, the doctors, the numbers of intelligent working men whom I know to be interested in the concerns of their country.' His reasons were that the politicians could not 'ensure the efficiency of our small army and the necessary supremacy of our navy without your support, sir, or madam, who chance to be reading this. It is in proportion as the nation refuses by practical voting to allow the question of our foreign policy to be made the shuttlecock of party that foreign statesmen are ready to give as we give.'[13]

As the facts overtook the fantasies, the new practice of forecasting began to acquire a modest methodology of its own. At the same time, as writers worked to transform their deductions about coming things into projections that would convince their audience, they came increasingly under the influence of the mass press. This was a crucial development, since the new journalism of the 1890s required writers (even admirals) to adopt a direct, factual and lively style of writing; and the circulation policies of the editors created a demand for original, informative and interesting material.

A most instructive example of this fruitful relationship between expert writer and general reader appears in the calculations that led to the publication of *The Great War of 189–*, which began in the January of 1892. This was specially written for the popular illustrated magazine *Black and White* by a consortium of senior officers: Rear-Admiral Sir Philip Colomb had earned an international reputation for his work on signalling systems for the new ironclad navy and he had written extensively on naval matters; Colonel Maurice was one of the most reputable military theorists of his day and he had distinguished himself as Professor of Military Art and History in the new Staff College; Captain F. N. Maude, a younger man than the others, was already well known for his publications on military affairs. The solitary civilian, Charles Lowe, was a former foreign editor of *The Times* and his thirteen years as the correspondent in Berlin for *The Times* made him well able to hold his own with the military. These eminent persons agreed to collaborate in an account of a future war, because the editor of *Black and White* believed that a forecast of 'the next great war' would interest his readers; and in his introduction to the serial the editor cheered the reader on with the news that 'this War will be fought under novel and surprising conditions. All facts seem to indicate that the coming conflict will be the bloodiest in history, and must involve the most momentous consequences to the whole world.'

The semi-fictional treatment of carefully calculated probabilities in *The Great War of 189–* was the point where forecasting began to take over from the older methods of the imaginary war story. In *The Great War of 189–* everything was subordinated to the demands of an exact and realistic projection. The story combined illustrations that were meant to create the illusion of photography with a variety of narrative styles that gave the impression of correspondents' reports direct from the battlefields. The account ran true to the political expectations of the 1890s in the forecast of the events that led to the outbreak of the war: an attempted assassination in the Balkans causes the Serbs to move against Bulgaria; the Austrians occupy Belgrade and the Russians begin to mobilize; France then declares war on Germany and British troops occupy Antwerp to prevent any advance through Belgium. By May most of the European countries are involved in the war. By December it is all over. Everything has gone with exceptional rapidity: divisions have on many occasions manoeuvred independently; infantry have advanced in massed battalions and formed square against cavalry.

On one occasion the authors came close to forecasting what was to happen later on in the First World War. They described a night attack by Russian troops who approached German positions with the aid of searchlights. When the Russians came within twenty yards of the enemy,

> ... their onward career was suddenly stopped short by some invisible barrier which made them crowd upon each other like penned cattle, passive targets for the bullets of our repeating rifles that rained upon them thick and fast as hail ... This barrier, which thus strangely stemmed the rush of their storming tide, was composed of fencing wire of several coils, strongly attached and impaled, which had been run along the front of our entrenched lines as an additional measure of defence.[14]

Up to this point the authors had been unusually perceptive; they had foreseen the use of barbed wire and they had accurately estimated that in the circumstances of the imagined battle the Russians could be expected to lose some 10,000 killed and wounded. Had they applied that deduction to the other engagements of their European campaign, they would have described a very different kind of war. Although Admiral Colomb and his military collaborators had tried to be as far-sighted as possible, they were unable to break with the habit of regarding military innovations as a means of speeding up methods of fighting that had not altered since the Battle of Waterloo.

There is a lesson for the futurologists of the twentieth century in the fact that the civilians did better than the military experts in forecasting the conditions and the consequences of a great European war. Because they were not burdened by the tactical doctrines of the staff colleges, they could pursue original lines of investigation and follow conjecture wherever it led them. As free agents, they were ready to consider the unthinkable and to employ the latest statistical techniques in their exploration of the future. It is not surprising, then, that the first major prediction in the new style was the work of an economist and banker – the Polish writer Ivan Bloch, who had spent twelve years in an exhaustive analysis of recent wars. His voluminous conclusions burst upon Europe in 1897 as a startling revelation for all but the military; and the title told the world that Bloch intended *The War of the Future in its Technical, Economic and Political Relations* to be a most comprehensive prediction.

The book was the first operational research investigation in the history of warfare. Bloch gathered most of his information from French and German documents on the war of 1870—ammunition supplies, rates of fire, use of weapons, deployment of troops, casualty lists—and he deduced that 'everybody will be entrenched in the next war. It will be a great war of entrenchments.' Written into the record of the future was a warning that the immense fire power of the vast conscript armies would lead to 'increased slaughter on so terrible a scale as to render it impossible to get troops to push the battle to a decisive issue'. Bloch had seen how technology had altered the scale of warfare: 'The war, instead of being a hand-to-hand contest in which the combatants measure their physical and moral superiority, will become a kind of stalemate, in which neither army being able to get at the other, both armies will be maintained in opposition to each other, threatening each other, but never able to deliver a final and decisive attack.'[15] Bloch predicted that it would become a war of attrition which would end in economic disruption, social disorder and even revolution.

This careful progression from demonstrated cause to probable effect was typical of the best investigatory thinking that established the practice of forecasting at the end of the nineteenth century. The development followed naturally on the universal experience of progress and on the application of new methods of social analysis; it was the last and most complex stage in the general examination of the future that began in the self-congratulatory visions of science fiction and the technological utopias, as well as in the more circumspect anticipations of the imaginary wars and the ideal states. The first forecasts were, in fact, the latest answers to the central questions of that age—What are we? Where are we going? What must we do? That they were a special means of informing the public was made clear to the readers of the *Illustrated London News* when the reviewer of Samuel Laing's *Problems of the Future* wrote of the merits of that book in 1889:

> A veteran public man, who has during forty years past, after gaining high mathematical honours at Cambridge, been constantly employed in Board of Trade official administration, or in the duties of a member of Parliament, or as Finance Minister of India, or as Chairman of the Brighton Railway Co., is not likely to be a dreamy idealist or a scholastic bigot. He has a vast acquired fund of practical common-sense, a trained faculty of induction and experimental reasoning, the habit of looking all round the different sides of a question, and then hitting the central point ... This instructive volume also contains several essays of a political character, on the huge military armaments of Europe, on taxation and finance, on the increase of population and the prospects of an adequate food-supply, concerning which the opinions of Mr. Laing should have considerable weight.[16]

This search for an equation of probabilities – for a law of socio-dynamics that would explain the shape of coming things – was a corporate activity in which all could give and receive according to the measure of their knowledge. In 1890, for example, Alfred Marshall published a most influential book, *Principles of Economics*, in which he argued that a Principle of Continuity guided the operations of human society in all ages and through all time. As Marshall was Professor of Political Economy in the University of Cambridge, he was no doubt expressing the received opinion of the learned when he asserted: 'The notion of continuity with regard to development is common to all modern schools of economic thought, whether the chief influences acting on them are those of biology, as represented by the writings of Herbert Spencer; or of history and philosophy, as represented by Hegel's *Philosophy of History*, and by the more recent ethico-historical studies on the Continent and elsewhere.'[17] The readers who persevered with the professor to the last chapter, 'The Influence of Progress on Value', were invited to rejoice in his conclusion that the great powers of technology were 'preparing the way for true self-government and united action by the whole people, not merely of a town such as Athens, Florence, or Bruges, but of a broad country, and even in some respects of the whole civilized world'. The state of future society was already apparent in the conditions of 1890: 'The diffusion of knowledge, the improvement of education, the growth of prudent habits among the masses of the people, and the opportunities which the new methods of business offer for the safe investment of small capitals – all these forces are telling on the side of the poorer classes as a whole relatively to the richer.'[18]

By 1890 everyman had become his own forecaster; and during the last ten years of the century the number of forecasts increased with exceptional speed throughout the periodical press, as the editors worked to satisfy the considerable interest in the changes to be expected in the twentieth century. Not a year went by without at least one article about the future finding a place in most of the periodicals. The record of the *Contemporary Review*, for instance, shows what was presumably the working of a consistent editorial policy, since a regular sequence of yearly articles began in December 1888 with thoughts about 'The Future of Food' and ended the century with appropriate considerations on 'The Social Future of England' in December 1900.

At the same time there was an equally unprecedented flood of books about the future, as writers in Europe and North America set out their forecasts for the twentieth century. Most of them showed the influence of recent developments in social theory – in anthropology, eugenics, and particularly in social science – and a special stimulus for the British interest in the future of the nation followed on the monumental survey of the London poor that Charles Booth began in 1889. Beatrice Webb, who saw what was happening, noted in her autobiography that the detailed descriptions to be found in the social survey reports 'became a characteristic feature of the publications of this period, whether newspapers or magazines, plays or novels, the reports of philanthropic organisations or the proceedings of learned societies'. Because the main tendency of sociological theory at that time was evolutionary and positivist, both popular and professional writing about the state of contemporary society was directed towards the future. 'The old dream of a bird's eye view of the past', as Beatrice Webb described it, 'and through it a glimpse into the future.'[19]

The sightings of the future agreed on the certainty of technological progress and at times agreed to differ about the probable condition of the world in the twentieth century. In one of the earliest German forecasts, *Deutschland im Jahre 2000* of 1891, Georg Ermann looked forward to an imperial Germany more united, more powerful and more prosperous than ever before. And in *The Control of the Tropics* of 1898 the so-called evolutionary philosopher Benjamin Kidd foretold that 'the events of real importance, those which are destined to shape and control the tendencies of history into the distant future, are those connected with the struggle for, and the occupation by the winning sections of the Western peoples of, those regions of the world where the white races can live permanently and work'.[20] Kidd ignored the contrary predictions in *National Life and Character: A Forecast* of 1893, where Charles Pearson came to a melancholy conclusion: 'It is now more than probable that our science, our civilisation, our great and real advance in the practice of government are only bringing us nearer to the day when the lower races will predominate in the world, when the higher races will lose their noblest elements, when we shall ask nothing from the day but to live, nor from the future but that we may not deteriorate.'[21] But cheerfulness kept on breaking in on the decline of the family and the decay of character; and Pearson was willing to accept the general view that 'the tendency of the age is to be hopeful, and it may be admitted that a great deal in the past history of the world encourages us not to despair of the future of humanity'. Indeed, the nations seemed to be advancing along the road to peace and co-operation:

> To the writer of these pages, what really seems most hopeful in the outlook for the future is the prospect that violent upheavals of society will be less and less attempted as the State appears to be the best expression of the wishes of the majority; and that some falling off in the energy and

> acquisitiveness, which are fostered by individualism, will be compensated by the growth of what we call patriotism, as each man identifies himself more and more with the needs and aspirations of his fellow-countrymen.[22]

Meanwhile, far away in St Petersburg the young Lenin had started on the revolutionary activities that led in 1895 to the foundation of the famous League of Struggle for the Emancipation of the Working Class. It is most unlikely that he ever read Charles Pearson on the future of the British people; and it is equally improbable that the political forecasters ever troubled to read what other writers were saying about the shape of society in the twentieth century. They gave their entire attention to their exclusive schemes for society in the conviction that the future would conform to their theories. Thus, in looking at the France of 1890, Edmond Darnaud found ample evidence for his belief in the anarchist associations of *La Société future*. And in 1894 a devoted follower of Theodor Hertzka, Ernest Udny, published a prospectus of the future, *The Freeland Colony*, on behalf of the British Freeland Association Expedition. Some sixty members were leaving to found a colony in the highlands of Kenya as their first move in changing the world: 'When Freeland has achieved its final purpose by the removal of poverty and material anxieties all over the world ... the nervous energy thus set free must follow other channels. We shall then be able to live (as men were fabled to have done in a past Golden Age) for the mere joy of living, and there will dawn upon the world an era of scientific progress and artistic achievement without a parallel in history.'[23] But anything that capitalism could do the socialists could do better. That was the theme of a book edited by Edward Carpenter in 1897, *Forecasts of the Coming Century*, in which ten eminent socialists set out the changes they expected to see in the future. Alfred Russel Wallace hoped to see self-supporting and co-operative labour-colonies spring up throughout the land; Tom Mann thought that the Trade Unions would grow rapidly in strength to the point when they would 'become part of the advancing democratic forces making for the Socialisation of industry'; and George Bernard Shaw wrote at length and with irony on 'The

Illusions of Socialism', asking his readers to clear their minds of the dramatic and religious illusions of their faith: 'By the illusion of Democracy, or government by everybody, we shall establish the most powerful bureaucracy ever known on the face of the earth, and finally get rid of popular election, trial by jury, and all the other makeshifts of a system in which no man can be trusted with power.'[24] Of all the forecasts in that book the only one that has any relevance to the late twentieth century came from Enid Stacy, who wrote on women's rights:

> Without venturing to prophesy or paint the ideal status of women in the various Utopias now floating about in so many reformers' minds, it will be safe to point out some of the developments which are not only historically probable but indispensable in the opinion of the believers in the powers and duties of women as citizens. No friends of the movement will be satisfied until women are free:
>
> 1. As individual women. The right to their own persons, and the power of deciding whether they will be mothers or not. The law actually denies even this elementary right to married women at present!
>
> 2. As wives. Perfect equality and reciprocity between husband and wife. This necessitates legal changes, notably as regards the Law of Divorce; *e.g.* whether the law be made laxer or more stringent, it must affect both sexes alike.
>
> 3. As mothers. Guardianship of their children on the same terms as in the case of fathers.
>
> 4. As citizens. The possession of the imperial as well as the local franchise, and full citizens' rights.
>
> 5. As workers. (*a*) For the present – whilst admitting the necessity of much regulation and many restrictions – to make as many of such regulations as possible applicable to both sexes. (*b*) Ultimately to obtain such a co-operative commonwealth as will ensure to each citizen, irrespective of sex, a choice of employment indicated by the results of education and only limited by individual capacity.[25]

For every Enid Stacy there were thousands who assumed that the twentieth century would be the nineteenth writ larger. For every Ivan Bloch there were battalions of staff officers, all ready for the expected war of movement, who would say later on with the Commander-in-Chief of the British Forces in France: 'No previous experience, no conclusion I had been able to draw from the campaigns in which I had taken part, or from a close study of the new conditions in which the war of today is waged, had led me to anticipate a war of positions.'[26] But accuracy in anticipation, it seems, depends upon the relationship between the predictor and the subject to be investigated. The best results come from those, who, like Ivan Bloch, have an expert capacity for analysing the factors at work in some limited area of human activity – technology, warfare, economics – but are not practitioners in that field; or, they come from those who, like Enid Stacy, have a moral commitment to specific social reforms that derive from the most profound human aspirations.

In fact, the course of predictive writing shows that, whenever a writer moves from the particular to the general, the more the prediction becomes a kind of social dramaturgy, an acting out of contemporary expectations between author and audience. Traditional beliefs, professional attitudes, customary roles, inherited symbols, sectional and national interests – these make it extraordinarily difficult for all but the most original of minds to break away from conventional patterns of thought and go voyaging on the unknown seas of the future. In consequence, it is a rare forecast that makes any allowance for the essential waywardness of human affairs and does not insist on a strict continuity between the self-evident present and the evidential future.

This habit of making a linear progression from the known to the knowable encourages a decided selectivity that gives the most-favoured nation the best of all conceivable futures and reserves for the less worthy a well-merited assortment of afflictions and disasters. In 1892, for example, Charles Richet introduced his forecasts in *Dans cent ans* with a long preamble on his methods of deciding on the probabilities of the future. He proposed to project 'the curve on the graph of progress forward from its base in past statistics to relatively significant probabilities, on the understanding that one accepts what is most likely – that is, the uniformity of phenomena.' Richet began, as so many do nowadays, with the known rates of increase in world population; and he concluded that, because Russia had a greater territorial area than that of the other European countries and because recent statistics had shown a fall in European birthrates, the Russians would treble their numbers by 1992. Compare his forecast of 340 million Russians with the recent United Nations Medium Variant projection of 334 million by 1995, and likewise compare his 750 million Europeans by 1992 with the projected 515 million by 1995. Since there would be a similar rate of growth in North America, Richet concluded that the two most powerful nations in 1992 would be the United States and Russia. It followed, therefore, that the Americans would seize Canada and the Russians would press forward into Asia; but there was always the possibility that the sleeping partner of the world, China, would wake to life, adopt Western armaments and 'rapidly chase the British from India, the French from Indo-China, and end as sole master of Asia'. In Africa the colonial powers – British, French, German, Portuguese – would still be there by the end of the twentieth century. A Gaullist severity, however, caused Richet to detach Egypt from the British. Egypt for the Egyptians – *voilà la solution de la question égyptienne!* To the French a more auspicious fate had reserved control of a powerful Franco-Arab empire that would stretch from the Mediterranean to the Sahara.

The rest of the forecasts followed the norm for that time: faster steamships, trains that travel at 60 miles an hour and flying machines that go a little faster than the trains. Industry will advance from discovery to discovery; men will find alternative sources of energy to replace the dwindling coal supplies; and 'the people of the twentieth century will have no fears for their food supplies. They will be better fed than we are, and they will have no more reason than us to worry about the future of their great-grandchildren. Even if we suppose that the human race will increase ten-fold, the land and the sea will be able to feed everyone: the prospects are completely reassuring.'[27]

Even more reassuring was the increasing interdependence of the nations that would advance the cause of world peace. Richet argued that by 1992 all standing armies could well have vanished, their place taken by an international tribunal. His reasons display the social drama of forecasting in a most vivid manner, since Richet follows the graph of technological development to a sentimental conclusion:

> Quick-firing rifles, monstrous artillery, improved shells, smokeless and noiseless gunpowder – these are so destructive that a great battle (such as there never will be, we hope) could cause the deaths of 300,000 men in a few hours. It is evident that the nations, no matter how unconcerned they may be at times when driven by a false pride, will draw back from before this fearful vision.[28]

It is a salutary comment on the millennial visions of *Dans cent ans* that unpredictable events were already preparing the course of history in the twentieth century. The deaths of millions had been decided in the lost childhood of Adolf Hitler; and the future of Russia waited on the young Joseph Stalin, who was then thinking of entering the theological seminary in Tiflis. The breakneck industrial development then going on in Japan was to lead to the destruction of the Russian fleet at the Battle of Tsushima; and the Russian defeat in the war with Japan was to be a major cause of the upheavals that brought about the Revolution of 1905.

But so powerful were the convictions of permanence in human institutions that the first century of industrial expansion closed with forecast after forecast that promised well for the world: E. Darnaud, *La Société future*, 1890; G. Ermann, *Deutschland im Jahre 2000*, 1891; J. and G. Simon, *La Femme au vingtième siècle*, 1892; A. Offermann, *Ueber die Zukunft der Gesellschaft*, 1893; Pierre Dronier, *La Navigation aérienne*, 1894; E. Clodd, *A Primer of Evolution*, 1895. And they went on pouring from the presses in Europe and in the United States year by year, as the forecasters set out their intimations of progress – with Josiah Strong on *The Twentieth Century City* in 1898, Gustave de Molinari on *Esquisse de l'organisation politique et économique de la société future* in 1899, Wells on *Anticipations* in 1900, George Sutherland on *Twentieth Century Inventions* in 1901, and Ernest Tarbouriech on *La Cité future* in 1902. And so on and so on, year by year up to the fatal August in 1914 when the outbreak of the First World War took H. G. Wells by surprise. On the evening of the declaration of war he began to write an article that gave the nation a slogan.[29] The title was 'The War that will end War', and he called on all 'to spread this idea, repeat this idea, and *impose upon this war* the idea that this war must end war'.

8

The Exploration of the Future

Long before the First World War destroyed the Wellsian hope of a peaceful, planetary society, H. G. Wells helped to bring in the twentieth century with a characteristic act of prediction. In a series of articles for the *Fortnightly Review* in 1900 he presented his *Anticipations* as 'a rough sketch of the coming time, a prospectus, as it were, of the joint undertaking of mankind in facing these impending years'. In the opening paragraph Wells claimed that his forecasts were a rare achievement, arguing that 'hitherto such forecasts have been presented almost invariably in the form of fiction'; and in a discreet footnote he again drew attention to the singularity of his speculations. 'Of quite serious forecasts and inductions', he wrote, 'the number is very small indeed.'

Wells was wrong; the number was already large. And had he told the whole story, he would no doubt have written that the discovery of the future, as he called it, had begun to attract the general interest in the 1890s. He could have added with complete truth that *Anticipations* was the first comprehensive and widely read survey of future developments in the short history of predictive writing. In its time it represented a peak in human self-awareness; it marked an important stage in the popular adaptation to those new directions in thought that Malthus and Condorcet were among the first to explore. Coming at the end of a long and prolific period of growth in futuristic literature, Wells in his day did more than any other writer since Malthus to promote an understanding of the future as a subject for rational inquiry.

The revelations of his science fiction stories and the persuasive clarity of his *Anticipations* provide the classic examples of the explorations of the future that had been going on ever since the last quarter of the eighteenth century. Because so many had toiled away at refining the various modes of describing the future, Wells was able to set his exceptional gifts of imagination to work in clearly defined areas of writing. Because he came to forecasting from a series of most original scientific romances, Wells knew how to popularize the idea of the future by presenting the drama of time-to-come as a miracle play in which progress and posterity play the parts of divine power and angelic chorus. In fact, his anticipations were a most accomplished self-confidence trick, since he began from the evolutionary base of expanding urban societies and with the assumption that right reason would always prevail in human affairs. In consequence his forecasts were – and were not – better than the rest of the field. His argument for English as the future world language lacked the statistical rigour of de Candolle; his failure to take account of calculable growth in populations led to a grotesque under-estimating of the role of Russia in the twentieth century; and despite the brilliant intuition of the land ironclads, the occasional and incidental projections of mechanical developments were inferior to those made by George Sutherland in *Twentieth Century Inventions*. The most remarkable aberration was that, although Wells made extensive use of Ivan Bloch's study of modern warfare and understood the changes that would come with the use of new weapons, he concluded that wars would be more rapid and that victory 'should be more easy in the future even than it has proved in the past'.

The main strength of *Anticipations* was the unusual scale of the enterprise. Wells was the first forecaster to consider the organization, social and political groupings, practices and beliefs that would follow on the constant technological development, continued expansion of education and increasing interrelatedness of the industrial nations. He expected cities to go on growing in size – a population of well over twenty millions in London and some forty millions in New York – and these would exist in vast urban regions, all held together by a network of railways, motorways, parcel delivery tubes and communication services. He thought that 'geographical contours, economic forces, the trend of invention and social development, point to a unification of all Western Europe.' The Little Englander in Wells made him cleave to his own, and he predicted a most advantageous union of the English-speaking peoples:

> A great federation of white English-speaking peoples, a federation having America north of Mexico as its central mass (a federation that may conceivably include Scandinavia) and its federal government will sustain a common fleet, and protect or dominate or actually administer most or all of the non-white states of the present British Empire, and in addition much of the South and Middle Pacific, the East and West Indies, the rest of America, and the larger part of black Africa.[1]

Within the major industrial countries a similar synthesis would emerge from the reform of schools and universities; and it could be that 'a confluent system of Trust-owned business organisms, and of Universities and re-organized military and naval services may presently discover an essential unity of purpose, presently begin thinking a literature, and behaving like a State.'[2] The educated and experienced professional men of the future would form the New Republic and they would in the fullness of time create the world state:

> And how will the New Republic treat the inferior races? How will it deal with the black? how will it deal with the yellow man? how will it tackle that alleged termite in the civilized woodwork, the Jew? Certainly not as races at all. It will aim to establish, and it will at last, though probably only after a second century has passed, establish a world-state with a common language and a common rule. All over the world its roads, its standards, its laws, and its apparatus of control will run. It will, I have said, make the multiplication of those who fall behind a certain standard of social efficiency unpleasant and difficult, and it will have cast aside any

> coddling laws to save adult men from themselves. It will tolerate no dark corners where the people of the Abyss may fester, no vast diffused slums of peasant proprietors, no stagnant plague-preserves.[3]

The main appeal of *Anticipations* was the swoop and the sweep of the narrative. Wells had orchestrated a symphonic rhapsody which opened with inductions from the present and closed with the enchanted music of everlasting progress. It was an inspired performance, for Wells was more prophet than predictor in the way he explained the necessary connexions between contemporary developments and the transformations of the coming years. It was a success story; Wells had created the world of the future in the image of his own experience. He had advanced by the power of intelligence and the gift of education from the chaotic confusions of early life in Atlas House to the stability and comforts of Spade House; and in like manner, by the same power of intelligence and the reform of education, mankind would go forward from an unsatisfactory present to the greater prosperity and political unity of the future.

Language, the style of writing and the presentation of ideas – these were the persuasive means of convincing the reader that the vision would come true. Wells obtained his effect by placing contemporary developments in an evolutionary context, by the frequent use of *primitive* to describe current practices, and most of all by the consistent manner in which he presented humanity as eternally on the move from a cruder past to a better future. Within this dynamic narrative system he developed an emollient style that allowed him to slip from debatable statements about the past to hypothetical claims for the future. To that he added a battery of powerful devices – an intimate conversational tone, direct appeals to the reader, tactical exclamations, vivid pictures of possible events, and physical metaphors that animated his explanation of evolving social organisms. He exploited a lilting, euphoric language whenever he wrote of the better things-to-come: 'Here about the great college and its big laboratories there will be men and women reasoning and studying; and here, where the homes thicken among the ripe gardens, one will hear the laughter of playing children, the singing of children in their schools, and see the little figures going to and fro amidst the trees and flowers.'[4]

Whenever Wells sought to turn the reader against whatever he disliked, he was careful to employ contemptuous and derisive phrases. He would contrast, for example, the brave and smiling Japanese with 'the pettifogging muddle of the English House of Commons'; or he would point to 'the tawdry futilities of army reform that occupy the War Office'; or he would claim that 'the headmasters entrusted with the education of the bulk of the influential men of the next decades are conspicuously second-rate men, forced and etiolated creatures, scholarship boys manured with annotated editions.' It is no wonder that, as soon as Beatrice and Sidney Webb had read *Anticipations*, they appeared riding very rapidly on bicycles from the direction of London to ask Wells to join and stimulate the Fabians.

The invitation is a striking example of the process of convergence then going on in every area of contemporary thinking about society and the future of society. By the end of the nineteenth century the old separateness between the professions and the public had gone. Civil servants worked openly to transform the state and admirals shared their hopes and fears with the nation in ways never known before. Political and social reformers—writers, trade unionists, colonial governors—had taken to collaborating in programmes that promised to change everything from the organization of cities to the treatment of the old and the sick. Again, for the first time in the long history of utopian literature, the old barriers between ideal theories and their social applications had vanished; and writers like Bellamy and Hertzka were able to attract devoted groups of followers who sought to turn their theories into realities.

In fact, the idea of the future had become so much a part of the general understanding of life in the great technological societies that fictional projections could at times decide the careers of inventors and could even affect the destiny of a nation. For proof there is the testimony of the pioneer of the modern submarine, Simon Lake, and of his contemporary who did so much for the development of aviation, Santos Dumont. The two men have explained in their memoirs how a reading of Jules Verne set them on their chosen paths of research and discovery. As has already been said, in 1865 Jules Verne had begun a new page in the history of science fiction with the first acceptable story of space travel, *From the Earth to the Moon*; and after reading that story a young Russian mathematician, Konstantin Tsiolkovsky, discovered the first stirrings of an interest in rocket dynamics. 'The first seeds of the idea', he wrote, 'were sown by that great, fantastic author, Jules Verne; he directed my thought along certain channels, then came a desire, and after that, the work of the mind.' By 1895 Tsiolkovsky was at work on a mathematical investigation of 'the possibilities of a jet ship as the most feasible means today for interplanetary travel outside the earth's atmosphere'. In 1903, after many hesitations by the editors of the *Nauchnoye Obozrenie*, the famous paper on 'The Exploration of Space with Reactive Devices' finally appeared in print.

This connexion between the imagined possibilities and the calculated practicalities of interplanetary navigation was typical of the way in which ideas about the future spread throughout the world. In 1896, for example, Theodor Herzl made the first moves in the Zionist campaign for an autonomous Jewish state with the publication of his pamphlet on *Der Judenstaat*;[5] and it was a telling indication of the close associations then existing between works of prophecy and prediction that in 1899 Herzl began writing a tale of the future about the foundation of a Jewish state in Palestine. When *Altneuland* appeared in 1902 the vision of a technological, democratic and religious Jewish nation had immense effect in shaping the image of the most desirable future for the Jewish people.

This progression from the analysis of community needs to the prophecy of their realization was a characteristic of the propagandists and persuaders who worked to impress upon the public their ideas about the future. To write about the future could be a work of rational prediction; it could with equal effectiveness be the product of a world-creating imagination that chose to picture the destiny of a people or of all mankind; and it could be a move in the fashionable game of finding the shape of things-to-come. Émile Zola, for example, spent the last five years of his life in writing the books of *The Four Gospels*; and in these he traced novel by novel the experiences of the children of the unfrocked Abbé Pierre Froment from the birth of Jean Froment in 1898 to the separation of Church and State about the year 1980. Zola established the setting for his stories by introducing occasional episodes and statements that indicate the social and technological advances of the twentieth century. In the first book, *Fruitfulness* of 1899, he foresees the great wave of human life flooding out from France to populate the world. The French colonists are already settled in the Niger valley. They are the promise of a new tomorrow:

> There was now no longer any mere question of increasing a family, of building up the country afresh, of repeopling France for the struggles of the future, the question was one of the expansion of humanity, of the re-claiming of deserts, of the peopling of the entire earth. After one's country came the earth; after one's family, one's nation, and then mankind. And what an invading flight, what a sudden outlook upon the world's immensity! All the freshness of the oceans, all the perfumes of the virgin continents, blended in a mighty gust like a breeze from the offing. Scarcely fifteen hundred million souls to-day are scattered through the few cultivated patches of the globe, and is that not indeed paltry, when the globe, ploughed from end to end, might nourish ten times that number?[6]

These fertile fantasies change in the second story, *Work* of 1901, to the social advances of the late twentieth century that follow on the last war in history. 'Ah! the last war, the last battle! It was so frightful that when it was over men for ever destroyed their swords and their guns ...'

> ... one-half of Europe rushed upon the other half, and other continents followed them, and fleets of ships battled on all the oceans for dominion over water and earth. Not a single nation was able to remain apart, in a state of neutrality, they all dragged one another forward ... And on all sides there was lightning, entire army corps disappeared amidst a clap of thunder. It was not necessary that the combatants should draw near or even see each other, their guns carried long miles, and threw shells which in exploding swept acres

> of ground bare, and asphyxiated and poisoned all around. Balloons also threw bombs from the very heavens, setting towns alight as they passed. Science had invented explosives and murderous engines which carried death over prodigious distances, and annihilated a whole community as an earthquake might have done ... And that was the last battle, to such a degree did horror freeze every heart when men awakened from that frightful intoxication, born of greed for dominion, lust for power; whilst the conviction came to all that war was no longer possible, since science in its almightiness was destined to be the sovereign creator of life, and not the artisan of destruction.[7]

About the turn of the century the comfortable theory of the utopian prophets held that a great war could be the prelude to a more perfect and most peaceful social order. This belief often appeared in the popular fiction of the time as the story of some future Edison whose inventions give him control of the world. In the January of 1893, for example, *Pearson's Weekly* began a serial story which entranced the readers of that magazine. The author was the enterprising journalist George Griffith, and in the 39 instalments of *The Angel of the Revolution* he worked through the contemporary formulas of compressed-air guns and fast aerial cruisers, of the plots of international anarchists and the war that would end all wars. By the end of the serial the forces of the International Brotherhood have defeated the armies of the Franco-Russian Coalition; they have proclaimed the Anglo-Saxon Federation of the World; they have abolished all armies and fortifications, and have established a final peace throughout the planet. These happy tales of the future continued for two decades, all working out their favoured variations on the great theme of war and peace: H. Lazarus, *The English Revolution of the Twentieth Century*, 1894; F. A. Fawkes, *Marmaduke, Emperor of Europe, Political and Social Reformer*, 1895; Wirt Gerrare, *The Warstock*, 1898; George Griffith, *The Great Pirate Syndicate*, 1899; Simon Newcomb, *His Wisdom, The Defender*, 1900. And so they continued their prophecies of conflict and concord until 1914, when H. G. Wells produced the last of them in *The World Set Free*, where an iron determination drove him to conjure the World State out of an atomic war.

These popular stories, like the more serious utopian prophecies of Émile Zola and H. G. Wells, followed out of the restatement of the idea of progress in the last decade of the nineteenth century. Many essays and books about the progress of society agreed on the proposition that the sciences had made possible all the extraordinary advances of the age and would go on with the great work of improving the conditions of life in the industrial nations. One ardent apostle of the doctrine of progress, Prince Kropotkin, stated this scientific credo for the readers of the *Nineteenth Century*:

> When we cast a glance upon the immense progress realised by all the exact sciences in the course of the nineteenth century, and when we closely examine the character of the conquests achieved by each of them, and the promise they contain for the future, we cannot but feel deeply impressed by the idea that mankind is entering a new era of progress.[8]

This conviction of the beneficent role of science was so powerful that at times it acquired something of the veneration once given to religion. For Somerset Maugham science was 'the consoler and healer of troubles'.[9] For Sir Leslie Stephen it was the repository of absolute truth, in which 'the primary axioms are fixed beyond the reach of scepticism' so that 'every new discovery fits into the old system, receiving and giving confirmation'.[10] Many thought that, as Wells put it, they were living 'in a period of adventurous and insurgent thought, in an intellectual spring unprecedented in the world's history'. Men like Osbert Sitwell came to maturity 'reflecting how wonderful it was to think that, with the growth of commerce and civilisation, mass captivities and executions were things of the rabid past, and that never again a man would be liable to persecution for his political or religious opinions.'[11]

For many Edwardians the past and with it all the customs and beliefs of the race had been largely discredited. The new writers – Wells, Shaw, Bennett, Galsworthy – saw the legacy of the past as something crude, outworn, dead. Indeed, for men like Wells and Shaw the past was unthinkable; it had all been a mistake, and mankind had to learn how to build a better world. A new future waited for them, as the young Osbert Sitwell realized, for they 'were being conducted by the benevolent popes of science into a Paradise ... of the most comfortably material kind: a Paradise where each man and woman even if no longer born with an immortal soul, could by means of such devices as false teeth and monkey glands have conferred on them a sort of animal and mechanic immortality of this world.' Expectations of this kind caused one enthusiastic prophet to break into galumphing verse:

Millennium! Millennium!
The wondrous world ordained to come!
A country world, as it were painted,
Of fruit and flowers in air untainted.
Where pregnant nature into beauty springs,
Yet lends to science still her soaring wings
To ever mount with purpose new,
Hidden mighty forces to subdue.

Wherein the ancient realm of might
Is quite subjected to the laws of right,
Where ambitious wars entirely cease,
And all the world is wrapped in peace.
Where health and vigour rule sublime
Suppressing weaknesses and crime.
The barren heath with culture tilled,
Life with universal love full filled.

Where plenty spreads her copious stall
With wholesome food enough for all.
With domestic bliss, of wife and child
Our leisure hours from care beguiled.
Where refined religions, soothing charm
Protects the soul from every harm.
Millennium! Millennium!
The wondrous world that is to come.[12]

The author was William Ford Stanley, a scientific instrument maker and inventor of merit, a frequent writer on scientific subjects, and the generous benefactor of educational institutions. In 1903 he published a most earnest utopia, *The Case of The. Fox, being his Prophecies under Hypnotism of the Period ending A.D. 1950*; and in it he described how a clairvoyant, Theodore Fox, had foreseen the course of progress in the twentieth century. His revelations were the standard prophecies of the Edwardian period: all the nations of the world are united in 11 great federations, from the States of Eastern Asia to the United States of Europe; wars and armies are things of the past; town-planning regulations have created an admirable environment in all great cities; the marriage laws of Europe control the health and well-being of every husband and wife; and a national health scheme operates in all the countries of Europe. The passages on religion–separation of Church and State–could have come direct from Émile Zola; and H. G. Wells could have written in more elegant language this account of the new sources of energy:

> As to energy, we have a large amount of tidal energy now conserved, as in the 'Wye Electric Storage Works'. We have also wave energy conserved, as in 'The Electric Wave Energy Company of Ilfracombe'. We have, of course, the energy of rainfall in rivers and mountain streams all over the world, now largely conserved through electricity; also that of direct sunshine, which we now focus in the centres of certain large valleys from glass reflectors producing intense heat by the direction of secondary reflections into limited areas. Also the energy of winds, conserved by our new large horizontal mills. We have also the heat energy of certain volcanoes, electrically conserved by modern appliances, besides the energy of coal and mineral oil, which was the most popular at the commencement of the century, but which has now become less economical than that from other sources.[13]

There had to be a reaction against this constant asserting of the primacy of the sciences; and the answer came one year later from G. K. Chesterton in the glorious history of *The Napoleon of Notting Hill*. He derided the progressive notion that bigness is best, that speed will be the salvation of society, that things will go on improving in the same way as they had been doing since the time of James Watt. By means of his prophetic allegory Chesterton stated what he was to go on saying for the rest of his days: that the human race can only save itself from disaster by returning to the local patriotism and universal ideas of the Middle Ages. The super-state was anathema to him; he felt that the administration must be so remote from the administered that it would frustrate its own purpose. Let human beings, he suggests, first seek to love and serve their own community, and then the well-being of nations and continents will follow.

Chesterton begins his holy war with the ironic observation that 'the way the prophets of the twentieth century went to work was this':

> They took something or other that was certainly going on in their time, and then said that it would go on more and more until something extraordinary happened. And very often they added that in some odd place that extraordinary thing had happened, and that it showed the signs of the times.
>
> Thus, for instance, there were Mr. H. G. Wells and others, who thought that science would take charge of the future; and just as the motor-car was quicker than the coach, so some lovely thing would be quicker than the motor-car; and so on for ever. And there arose from their ashes Dr. Quilp, who said that a man could be sent on his machine so fast round the world that he could keep up a long chatty conversation in some old-world village by saying a word of a sentence each time he came round.[14]

Chesterton goes on with the joke, saying that his future world is not at all like the imaginings of Wells, Edward Carpenter, Cecil Rhodes, or Sidney Webb. England has become a static, spiritless utopia, because the nation no longer cares strongly enough about anything: 'England was now practically a despotism, but not a hereditary one. Some one in the official class was made King.' Then everything changes when Auberon Quinn, the newly elected King of England, orders the London boroughs to revert to the splendour of free medieval cities, with all the glories of heralds, halberdiers, city walls and tocsins.

Of course, Chesterton had to have his war – for the fun of the boisterous sword-play and to make his point that every man must fight for 'the place where he had the Eden of childhood and the short heaven of first love'. The London boroughs go to war. Although Notting Hill may be the weakest of them all, in fact the Boer Republic of London, it represents all that Chesterton thought dear and sacred in life. So the trumpets sound, and Auberon Quinn goes into battle. As the reviewer in *The Academy* observed with complete truth, 'Mr Chesterton, who delights in turning things topsy-turvey, makes an ideal state the protoplasm of a civilised community.'

Six months after the appearance of Chesterton's heroic allegory in favour of civic courage and the civilized community H. G. Wells produced excellent evidence for the Chestertonian case against 'the prophets of the twentieth century' with his extravagant parable of change and development in *The Food of the Gods*. The story represented everything Chesterton detested in the Wellsian ideal state, since the giants in Wells's story are there to prove that bigness is best and that the permanent drive in human beings is towards the 'growth that goes on for ever. Tomorrow, whether we live or die, growth will conquer through us. That is the law of the spirit for evermore.' The workings of that law was the subject of Wells's next book, *A Modern Utopia*, which came out in 1905.

The acclamations that greeted *A Modern Utopia* were a sign that Wells had found yet another way of writing about the future of society that coincided with the mood and interests of his contemporaries. In fact, Wells had produced the most up-to-date utopia in the history of the genre. It was entirely in keeping with the contemporary demand for the more efficient management of society, since Wells started from the premise that 'were our political and social and moral devices only as well contrived to their end as a linotype machine, an antiseptic operating plant, or an electric tram-car, there need now at the present moment be no appreciable toil in the world, and only the smallest fraction of the pain, the fear, and the anxiety that now make human life so doubtful in its value'.

Once again Wells had come well prepared for a new task and a new role. In his earlier work he had gone from one imaginary society to another – from the hunters and the hunted in *The Time Machine* and the even more dreadful creatures in *The Island of Doctor Moreau* to the intricate lunar hierarchies of *The First Men in the Moon* and the opposing groups in *When the Sleeper Wakes*. From these imaginative schematizations Wells turned with equal facility to the very different circumstances of the social novels. His intention in *The Wheels of Chance*, *Love and Mr Lewisham* and especially in *Kipps* was to show how a neglectful society had sinned against the less fortunate. Hoopdriver, Lewisham and Kipps were, said Wells, 'all personalities thwarted and crippled by the defects of our contemporary civilisation'. This was the point that the Liberal politician Charles Frederick Masterman stressed in his lengthy review of *Kipps* in the *Daily News* for October 25th, 1905:

> Those familiar with *Mankind in the Making*, and the violence of its attack upon so much modern progress which is tolerated or applauded, will find in *Kipps* a kind of materialisation in fiction of that social impeachment ...
>
> All the mordant power of Mr. Wells's revolt against the mess which men and women are making of their world, against the failure of a life which has attained comfort but no inner serenity or passion or large and intelligible purpose of being is woven into his picture of the struggles of Kipps to attain a footing in these regions of social advancement. He is compelled to do the things he hates rather than the things he desires. He is driven to construct a gaunt and hideous villa, to live up to his station, instead of the little home of his dreams. Everywhere he is hedged in, before and behind, by the spiked and barren branches of aimless, conventional, respectable things.

Shortly after Wells finished *Kipps* in May 1904 he began work on *A Modern Utopia* 'where all that is tangled and confused in human affairs has been unravelled and made straight'. His scheme for a faultlessly administered and peaceful world-state was the consummation of the Baconian tradition. The inhabitants of the planet far beyond Sirius have created their New Atlantis by the powers of science and the light of education.

For Wells, as for Bacon, knowledge was the key to right action; and because 'the leading principle of the Utopian religion is the repudiation of the doctrine of original sin', it followed that an ideal commonwealth could only contain ideal human beings. The social arrangements would, therefore, decide the moral attitudes. That attractive proposition was part of the Wellsian formula for the just society that had its beginnings in his life-long indignation over the unhappy experiences of his adolescence:

> I have heard other people who have had similar experiences to mine tell of the thirst for knowledge they experienced. I suppose I had that thirst in good measure, but far stronger was my anger at the paltry sham of an education that had been fobbed off upon me; angry resentment also at the dismal negligence of the social and religious organisations responsible for me, that had allowed me to be thrust into the hopeless drudgery of a shop, ignorant, misinformed, undernourished and physically under-developed, without warning and without guidance, at the age of thirteen. To sink or swim. I was too young to make allowances for the people who were exploiting and stifling me. I did not realise that they were charming people really, if a little too self-satisfied and indolent ... But I did not discriminate about their responsibility. I hated them as only the young can hate, and it gave me energy to struggle, and I set about struggling for knowledge. I was bitterly determined to see my world clearer and truer, before it was too late.[15]

The anger and resentment explain the violent openings of so many of the Wellsian utopias. Wells usually took the opportunity to give mankind a fresh start by destroying the universe of established things. In *The Food of the Gods* the world drifts into chaos, because the old order was incapable of dealing with the consequences of an extraordinary scientific discovery. Again, in *The World Set Free* the coming of atomic energy and the outbreak of an atomic war lead to the same catastrophic results. 'What else can happen,' asked Frederick Barnet, 'when men use science and every new thing that science gives and all their available intelligence and energy to manufacture wealth and appliances, and leave government and education to the rusting traditions of hundreds of years ago?'[16]

These failures in foresight always brought out the angry reformer in Wells. In *Anticipations* he damned the inventors of the first steam locomotives as men of insufficient faith, because they had not thought out a better means of transportation than their railway systems; and he protested with equal vehemence against those failures in the sensible organization of human life that led to inefficient methods in the building, heating and cleaning of houses. *Chaos, muddle, failure, preventable, primitive, confusion, stupidity* were his more favoured words of condemnation; and he commended his own ideas to the reader by calling them *efficient, self-evident, functional, progressive, scientific, rational.* Like the narrator of that miraculous utopia, *In the Days of the Comet,* Wells felt that he lived in a world of preventable disorder, preventable diseases, and preventable pain. 'Here were we British,' the narrator cried out in indignation, 'forty-one millions of people, in a state of almost indescribably aimless economic and moral muddle that we had neither the courage, the energy, nor the intelligence to improve, that most of us had hardly the courage to think about, and with our affairs hopelessly entangled with the entirely different confusions of three hundred and fifty million other persons scattered over the globe.'[17]

This nagging, persistent urge to set the world to rights began for Wells some three months before his fourteenth birthday, when he was plunged into the tribulations of finding a way for himself through the commercial jungle; and the first vivid realization of the essential disparities in human life came to him in the Rodgers & Denyer drapery emporium, as he looked across from his station at the cash-desk and saw on the other side of the High Street the entrance to Windsor Castle. Twenty-five years later the young Bertie Wells had become one of the major writers of the day. In the opinion of George Orwell it was doubtful 'whether anyone who was writing books between 1900 and 1920, at any rate in the English language, influenced the young so much'. And Compton Mackenzie, writing in 1933, thought that Wells had exercised 'a more profound and a more extensive influence upon English life and upon English literature than any other writer of the twentieth century'. He had a vast following, as André Gide noted, recruited in all countries and from all social classes. He offered his readers more than any other writer—a seemingly inexhaustible variety of books that began with brilliant stories about the role of science in human affairs and led on to beguiling visions of the World State. And he worked on his readers by telling them that the past was no more than the prelude to a better future, by proclaiming the infallibility of the scientific method and by presenting his own very decided ideas on the great questions of the day—on everything from marriage and the role of women to the management of industrial society.

Throughout three decades of writing about the future, from *Anticipations* to *The Shape of Things to Come*, Wells created a unique body of promissory literature which was evolutionary in its explanation of social development and revolutionary in its forecasting of a coherent, co-operative planetary society. His peculiar power as the magus of a mechanistic age sprang from his constantly repeated conviction that the forces of progress worked ceaselessly for mankind; and the special charm he had for his readers owed much to the artful ways in which he matched style to substance in his prophecies. The magic of this gift is most evident in *A Modern Utopia* where Wells combined the allurements of the high predictive manner with the ruminative philosophizing of the older ideal states. It remains the most important utopia of the twentieth century, because it made the most complete act of faith in the idea of progress, and because by a cruel irony of fiction it became the model from which the dominant dystopias of the last fifty years have taken their mark.

In *A Modern Utopia* Wells produced the poor boy's New Atlantis; he transformed the remembered deficiencies of his early days into the everlasting sufficiency of the universal welfare state. It was both a dream and a demonstration. By employing another of his innovations, the device of a parallel universe, Wells was able to show what our world could become, if only the human race had wit enough to carry through the re-organization and re-education of all peoples on earth. The style worked with the theme like the piston in a high-pressure cylinder. Wells in the role of narrator was assured, emphatic, inquiring, omniscient, laudatory and properly disparaging as the occasion required; and to complete the general strategy he made the botanist, the narrator's companion, exhibit the most deplorable human characteristics – petulant, prejudiced, exclaiming that he could not live in utopia if there were not to be any dogs, protesting that he would not like his daughter to marry a Chinaman or a negro. From the start the reader is involved in the narration as explorer and judge; and he soon discovers that Wells refers constantly to the social mechanisms that keep the wheels of the great society turning. He writes about the impersonal energy of the citizens, the apparatus of existence, the general machinery of the state, and he says of his perfect planet that, 'compared with our world, it is like a well-oiled engine beside a scrap heap'. It is an urban, progressive, uniform and managed world in which there can be no place for such unmanageable creatures as Mr Kipps, Mr Polly, and George Ponderevo. For most of the book the design follows the usual system of the ideal states – universal peace, a world language, research centres humming with activity, the state ownership of land and of all sources of energy. The healthy, happy citizens accept their grading into any one of the four main classes. They are content with their station in life, content to live in their well-planned houses, content to follow the directions of their self-elected guardians, the Samurai.

In this Wells followed Plato in giving the direction of his ideal state to a ruling class chosen from the best and wisest citizens, who are prudent and self-controlled and dedicated to the service of mankind. The Samurai represent an exercise in social engineering. They are the managers of the World State; they look like Knights Templars and follow an austere rule of life; and they fulfil all the specifications of the Fabian scheme for the efficient administration of an industrial society. The Samurai are the crucial test of the Wellsian utopia, as G. K. Chesterton was the first to point out. Had Wells begun at the beginning – that is, with himself – Chesterton thought he would have discovered how 'a permanent possibility of selfishness arises from the mere fact of having a self, and not from any accidents of education or ill-treatment. And the weakness of all Utopias is this, that they take the greatest difficulty of man and assume it to be overcome, and then give an elaborate account of the overcoming of the smaller ones.'[18]

Nevertheless, the Wellsian scheme for the self-perfecting society at once found devoted adherents, and some of these united in 1906 to establish the Samurai Press at Ranworth Hall near Norwich. One of the founders was Harold Monro, who gained valuable experience of publishing from his work for the Press; and in a way no one could have foreseen this led to the verse anthologies Monro began to produce with Edward Marsh in 1912 that put the mark of Georgian Poetry on English literature. In 1907, however, Monro was committed to changing the world on the lines laid down in *A Modern Utopia* and he collaborated in writing the *Proposals for a Voluntary Nobility*, one of the most pretentious declarations in the history of utopian literature. The tract opened with the assertion that 'great men in all ages have set a standard of life for their successors ... The measure of any given moment in history is the ideals of its great men, its philosophers, its prophets, its poets, those who create, as cosmic energy creates, new excellence.'

Thus, the line of human progress runs from Plato and Pericles to the clearest-sighted idealist in England, H. G. Wells. The propagators of the Samurai ethic, like the angels of a New Jerusalem, trumpeted the good news that 'this generation has witnessed a marvellous change sweep over the world, a change that has almost incredibly unified the methods of seeking and in the end it must be of attaining what is now the ideal. Science has shown the possibilities of conscious evolution. Blind groping is no longer the characteristic of seekers for excellence and truth.'[19] Since Wells had revealed the way to a higher order of human existence, all that remained was for right-minded men (women only received the call in a footnote) to come together in a Voluntary Nobility 'in order that they may accomplish the work of Here and Now and leave the world better than they found it'. The goal of the volunteer nobles was nothing less than the spreading of the Wellsian gospel for the perfecting of the world:

> In these proposals there has been a general tendency to disregard the relation of a Voluntary Nobility to existing institutions, because it is clear that it must not stand in any sort of relation to them. What advantage it can gain from extraneous example or experience it will naturally take, but it will set its own standard and it will be self-contained. It will not set out to help sinking ships, but, in steering its individual course over the vast untravelled ocean, it will avoid colliding with them.
>
> In its very essence it must be progressive and kinetic. It must be entirely free from tradition. Quietly and unobtrusively it must move with but always a little above the level of its age. It must march in the forefront of progress and efficiency.[20]

The expectation of continued progress and the demand for higher levels of social efficiency were the beginning and the end of most Edwardian explorations of the future. As the proposals for a Samurai Nobility show, the utopias and the forecasts agreed that it would be possible to direct society towards the better life: that is, towards a more organized and, therefore, more contented way of living. So, the habit of expectation flourished more than ever before, stimulated by a hothouse atmosphere of prophecies and predictions that represented the onward movement of mankind as a uniform advance in social behaviour and technological capacity. Scientists and popular writers joined with all manner of social theorists in describing the changes that would follow in the twentieth century. The most enthusiastic forecasts appeared in the periodical press, especially in the illustrated magazines, which had extended the original practice of the 1890s and made articles about the future a part of their regular offerings to the public. And here the more mechanical and uncritical applications of the doctrine of progress promoted the belief that, as Hudson Maxim had it, all things were leading to 'Man's Machine-Made Millennium'. In an introductory note to Maxim's forecast the editor of the *Cosmopolitan Magazine* summed up the delights in store for humanity:

> The discovery of a radio-motor, says Mr. Maxim, will make

> power so cheap that none will work save for recreation; crystallization of fertilizer out of the atmosphere will make the earth so prolific that farming will be a pastime; disinfectant solutions forced through the body will exterminate all germs, and diseases will be eliminated; life insurance companies will become simply accident insurance companies, and man's life will run its allotted span; criminals will no longer be imprisoned, but will be segregated in a great reservation where they will live out their lives, the right to propagate their kind denied them, thus eventually cleansing the world of its criminal element; the mastery of the air will liberate mankind from the limitations of navigable rivers and railroad tracts; gold will be so common that it will be used for rifle bullets; diamonds as big as the Kohinoor will be made for a dollar, and the city of the future will not be a collection of buildings, but one vast arcaded building with its subdivisions carefully allotted for the needs of its inhabitants.[21]

The article followed along the track, well-worn by 1908, which H. G. Wells and George Sutherland had first explored about the turn of the century; for the exponents of the new informational journalism made it their constant duty and practice, as Sutherland explained in 1901, 'to take note of the advance of inventive science as applied to industrial improvement – to watch it as an organic growth, not only from a philosophical, but also from a practical, point of view'. Thus, it was entirely in keeping with the policies of contemporary journalism for the editor of *Harper's Magazine* to contract with Frederick Soddy for a special article on 'The Energy of Radium'. Soddy had worked with Ernest Rutherford at McGill University in the famous experiments that developed into the papers on the disintegration theory of radio-activity; and about 1903 an understanding of the potentialities locked up within the atom spread from the laboratories to the press. One reviewer in 1904 ended a report on the latest ideas about radio-activity with the disquieting suggestion of Professor Rutherford that, 'could a proper detonator be discovered, an explosive wave of atomic disintegration might be started through all matter which would transmute the whole mass of the globe into helium or similar gases, and, in very truth, leave not one stone upon another. Such a speculation is, of course, only a nightmare dream of the scientific imagination; but it serves to show the illimitable avenues of thought opened by the study of radio-activity.'[22]

The news about the atom continued to spread through the periodical press, and in the December of 1909 Frederick Soddy told the readers of *Harper's Magazine* that atomic energy would solve the twin problems of diminishing fuel supplies and increasing demands for energy: 'Sooner or later man must gain command of the newly recognized internal sources of energy in matter and control them for his own purposes, or he must lose much of the dominance he has already attained. On the other hand, if he succeeded in tapping these primary sources of energy, the future would bear as little relation to the past as the life of a dragon-fly does to that of its aquatic prototypes.'

The biological analogy sprang naturally from a state of mind that decided most of the assumptions made about the future in the years before the First World War. In that time of unavoidable innocence it was axiomatic that humanity would move for ever into the future like the pioneer wagons rolling inexorably towards the new frontier. In the stock phrases of the day society was to go on growing, evolving, expanding, developing and progressing as it had been doing ever since the invention of the steam-engine. For the enthusiasts of progress, like H. W. Hillman, the American author of *Looking Forward: the Phenomenal Progress of Electricity in 1912*, the reason for writing about coming changes was 'to present to the people the many evidences of other new conditions destined to greatly favor the domestic, social and industrial relations of the people in the pursuit of their daily vocations'.[23] For the more reflective there were good reasons for thinking that calculated actions must lead to predicted results. That was the belief of the eminent Egyptologist Flinders Petrie who turned from the archaic past in 1907 to set out his ideas about the future of society in *Janus in Modern Life*; and there he opened with the popular theorem that all peoples and all times are linked in a dynamic and progressive system: 'Every step of the past has been a present, living, urgent, imperative, to the whole world; and every such present has been entirely conditioned by its past, just as the future to us is conditioned by our present.'

For some of the more eager world-changers this search to discover future probabilities imposed the conclusion that science alone could take the guesswork out of prediction. Half a century before the appearance of the first journals of futurology some searchers after things-to-come had moved on from the position Wells established in *Anticipations* and had begun to work out a more systematic means of predicting the social changes that would take place during the twentieth century. Karl Pearson, the pioneer of biometrics and propagandist of eugenics, thought that it would be possible to make forecasting into a science. In his study of *Social Problems: their treatment, past, present and future* Pearson argued that the biological sciences would enable society to comprehend and control the future: 'The sociology of the future – nay, the very science of history in the future – will be a biological science. Human society has developed under those factors of environment, tradition, and heredity from the herd to the civilized nation. Can we learn the laws of that progress? Can we interpret those laws so as to assist future progress? Can we aid man to develop socially with less friction than in the past?'[24] The questions expected affirmative answers, since the entirety of progressive and predictive writing depended on the belief that the discovery of the future would follow from the investigation of significant patterns of development just as the knowledge of world geography had followed on the voyages of exploration. One observer thought that the recent growth in forecasting was the response of science to the deficiencies of utopian fiction. In his survey of 'Forecasts of Tomorrow' in the *Quarterly Review* for July 1908, William Barry argued that utopian literature had become

> ... a stock department in libraries, and has of late flourished with an abundance which may remind us of the pamphleteering that went on before the French Revolution. At least one hundred works in this kind have been circulated since Bellamy's 'Looking Backward' gave to its pages a Socialist colouring. But these are mainly fiction; and fiction, however

> effective as propaganda, will not satisfy the demand, thanks to which speculation concerning the future of civilized mankind is now rife. Hence a more scientific and serious method has given rise to publications which, whether founded or not in statistics, aim at reaching first principles, and if they end in prophecy, start with induction from present facts.[25]

But what could a more scientific method discover about the future when so many things were happening for the first time? Looking back, it is now possible to see what none could then foresee – that the expansionist policies of the new German Reich would lead to the first great technological war in human history, to slaughter on an unimagined scale, to the collapse of government in many countries, and to the emergence of despotisms which would deny all the hopes of predictive literature for the progress of mankind. Already in 1900, when Wells was writing so hopefully about the future, the second Navy Law had committed the Germans to a twenty-year programme for the construction of a great navy; and as the German battleships came down the slipways year after year, the portents grew less and less propitious. The Tangier incident of 1905 was an indication of the aggressive intentions of German policy and the Agadir crisis of 1911 made the prospect of a European war seem even more likely.

In that year a young German schoolmaster had retired on a small private income to study history and philosophy in Munich. As Oswald Spengler pondered the implications of the Moroccan crisis, a flash of intuition told him that a world war was coming and that this would be '*a historical change of phase* occurring within a great historical organism ... at the point preordained for it hundreds of years ago'. In the following year Spengler came upon a book about the collapse of the ancient world, and musing upon that calamity he convinced himself that contemporary civilization was entering upon a comparable series of catastrophes. The inspiration gave him the subject and the title for his apocalyptic work on *The Decline of the West*, and for five years he laboured to show that the idea of progress could not be the key to history. In a most appropriate way the scholar who was to be the first prophet of the post-war world had appeared in the last days of imperial Germany; and as he orchestrated his prodigious theme, he grew ever more convinced that the end of the Europe he knew was near. He dedicated himself with renewed determination to 'the venture of predetermining history, of following the still untravelled stages in the destiny of a Culture, and specifically of the only Culture of our time and on our planet which is actually in the phase of fulfilment – the West-European-American.'[26]

> There is not anything more wonderful
> Than a great people moving towards the deep
> Of an unguessed and unfeared future; nor
> Is aught so dear of all held dear before
> As the new passion stirring in their veins
> When the destroying dragon wakes from sleep.
>
> Happy is England now, as never yet!
> And though the sorrows of the slow days fret
> Her faithfullest children, grief itself is proud.
> Ev'n the warm beauty of this spring and summer
> That turns to bitterness turns then to gladness
> Since for this England the beloved ones died.

In that poem, 'Happy is England Now', written on the outbreak of the Great War, John Freeman composed a fitting epitaph for the idea of the future. He was not to know that out of the 'unfeared future' would come the most fearful surprises for the nations of Europe. The last irony was that after fifty years of prophesying and forecasting the Europeans believed that they were entering on an old-style war of rapid movement and small casualty lists. As they said in the August of 1914, it would all be over by Christmas.

PART THREE

From Millennium to Millennium

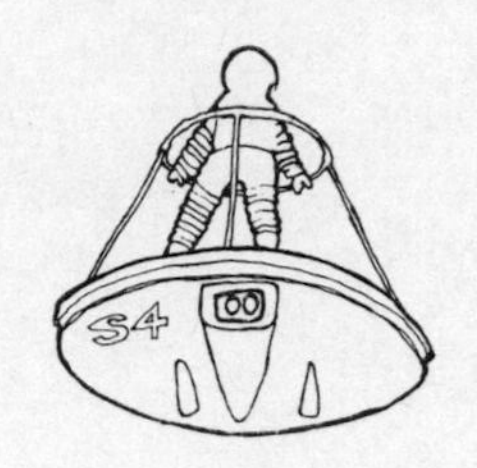

9

From Bad to Worse

The millions who went with songs to battle in 1914 expected to play their part in a short and glorious campaign. One of them, the historian Duff Cooper, thought that he was volunteering for an old-style war, because his reading of history told him that previous wars had never 'interfered very much or for very long with the civilian population'. Like almost all his contemporaries, said Duff Cooper, he 'lacked the imagination to conceive that this war was likely to differ to any great extent from the others'.[1] As the poet Herbert Read explained, this failure in anticipation followed from the fact that 'war still appealed to the imagination'.[2] The ideal of voluntary service in time of war was still part of the high style of a society that had failed to foresee how the new means of warfare would change the conduct of war.

Before the experience of the trenches changed his ideas, Siegfried Sassoon had believed that realism and war poetry did not go together. He was content to write in the traditional way about 'the woeful crimson of men slain';[3] and his contemporary Rupert Brooke had expressed this customary attitude to war in a poem that opened with the conventional, literary line: 'Now, God be thanked Who has matched us with His hour.' The poets welcomed the outbreak of war with archaic words and traditional sentiments that had not changed in a hundred years; and in this they were true to themselves and to the expectations of their nations.

At first, after all the forecasts and all the tales of the next great war, the imagination could not respond to the unexpected consequences of the new military technologies. The flood of futuristic literature shrank almost to nothing. As the fearful realities of barbed wire, poison gas, machine-guns and quick-firing artillery burst upon the astounded nations, only the front-line poets could find words to describe 'the monstrous anger of the guns'. In Wilfred Owen's phrase the poetry was in the pity; and one young poet, Charles Sorley, who was killed in action before Loos in his twentieth year, saw darkly how false ideas about the future had entrapped the combatants in a great conflict. He dedicated his poem 'To Germany':

> You are blind like us. Your hurt no man designed,
> And no man claimed the conquest of your land.
> But gropers both through fields of thought confined
> We stumble and we do not understand.
> You saw only your future bigly planned,
> And we, the tapering paths of our own mind,
> And in each other's dearest ways we stand,
> And hiss and hate. And the blind fight the blind.

The battles raged along the line of trenches from the Swiss frontier to the Channel coast, and out of the daily need to anticipate enemy actions came the beginnings of modern practice in technological forecasting. In a hesitant and often haphazard fashion the governments, scientists and industrialists collaborated in devising means and measures to meet the unparalleled exigencies of technological warfare. After the German armies had fired off most of the nitrate stocks during the invasion of France, the Raw Materials Department summoned the eminent German chemist Fritz Haber to make up the deficiency in nitrogen fertilizers; and after the failure of the German offensive on the Marne the *Waffenbeschaffungsamt* asked Haber to investigate the possibilities of gas warfare. Haber did his duty by providing chlorine gas and a serviceable gas mask. When the first clouds of poison gas rolled over the French positions to the north of the Ypres salient on April 22nd, 1916, the Allies at once set their scientists and military engineers to work on means of dealing with this new method of fighting.

The lesson of poison gas, submarines and aeroplanes for

governments at war was that, since they had the future of the nation in their keeping, it would be wise to find some ways-and-means system of dealing with the unprecedented problems of a technological society. In the United Kingdom the Department of Scientific and Industrial Research began to function in 1915, and in 1916 the Americans established their National Research Council. These innovations were a political and scientific acknowledgment that government had to plan and prepare for changes on a scale never experienced before. And so, out of these beginnings and out of the renewed production of futuristic writing in the 1920s there came new practices in forecasting.

The immense changes that followed on the First World War sent tremors of fear and anxiety through the old complacencies of futuristic literature. One immediate effect was a profound difference of opinion between the foreseers and the forecasters, who ceased to work from the same set of assumptions. A new generation of prophets tore up the agreement, implicit in most of the utopias and in almost all the forecasts, that more technology and more organization must lead to a better society. The once parallel streams of prophecy and prediction diverged. The forecasters limited themselves to looking for significant patterns of development, and the writers of fiction turned from celebrating the powers of man over nature to composing ominous, admonitory myths about the dangers that confront any technological civilization. Utopia became dystopia, and the once self-confident accounts of future wars changed to fearful visions of a planetary conflict in which poison gas, or giant bombing planes, or biological weapons wipe out the human race.

One of the first post-war tales of the future foretold how human folly would put an end to civilization. In 1920 Edward Shanks introduced the theme of the desolation-to-come in *The People of the Ruins*, where he shows that a succession of crippling wars have killed off most of the population of Europe. The survivors are little better than barbarians. They live in chronic ignorance and perpetual fear. London has fallen into ruins, and 'the wilderness that had been Regent's Park was a singular and striking reminder of the time when London was a great and populous city. Every stage of desolation and decay was to be seen in that appalling tract, which had lost the trimness and prosperity of its flourishing period without acquiring the solemn and awful aspect of nobler ruins.'[4]

The tales of ruin and desolation soon developed their characteristic style of narrative; and this has carried the same uncomfortable message for these last 50 years – from *The People of the Ruins* in 1920 to Aldous Huxley's gloomy version of the nuclear disaster story in *Ape and Essence* in 1949, to Walter Miller's vision of an end for an incorrigible mankind in *A Canticle for Leibowitz* in 1955, to Edmund Cooper's account of the Martian expedition that discovers survivors on Earth in *The Last Continent* of 1970. From the earliest to the most recent of these stories the survivors tell the same tale of human failure. They repeat their histories of the great disaster; they reveal their degradation by showing the ruins of London in *The People of the Ruins* or the wastelands of California after the Third World War in *Ape and Essence*.

The persistence of the disaster stories and the marked uniformity of the narratives indicate the vigour of a favourite twentieth-century nightmare – the fear that the human race has an unlimited capacity for self-destruction. There is, for example, no great difference between the themes and views of Aldous Huxley in *Ape and Essence* and those of Cicely Hamilton in *Lest Ye Die* in 1922. Both start from the experience of the last war, and for Cicely Hamilton this leads to a prophecy of the future wars that will wipe out contemporary civilization. Destruction, starvation and disease have obliterated the recognizable world, and 'that which had once been a people, and an administrative whole, was relapsing into a tribal separatism, the last barrier against nomadic anarchy'. Things have fallen apart for lack of moral sense:

> We were not civilised – it was only our habits that were civilised; but we thought that they were flesh of our flesh and bone of our bone. Underneath, the beast in us was always there ... We dressed ourselves and taught ourselves the little politenesses and ceremonies which made it easy to forget that we were brutes in our hearts; we never faced our own possibilities of evil and beastliness.[5]

The ending of the old life is complete. The narrator dwells at length on the future primitive society in which the vow of loyalty to the tribe is the supreme law and men are graded according to their skills. Because the machine was the cause of so many evils, these primitives of the future nourish a great hatred of all things mechanical. One of the most solemn moments in the new life is the taking of the 'Compulsory Vow of Ignorance'. This forbids all prying into the secrets of nature because of the disasters that came from these inquiries in the past.

A variant on the disaster story was the salvation myth. This relates how a man of genius, usually a scientist, saves a remnant of humanity and lays the foundations for a better order of existence. By virtue of this stratagem a writer can turn hell into heaven. He can punish the worst failings of society with the almost total obliteration of civilization, and thereafter he can show how men should live by building an ideal state on the ruins of the bad old world. The story that established this classic stereotype appeared in 1923 – *Nordenholt's Million* by J. J. Connington who was A. W. Stewart, professor of chemistry in Queen's University, Belfast. His tale of redemption begins with the scientific nightmare of *Bacterium diazotans* which creates a world famine by removing the nitrogenous content of the soil. In the face of this disaster the politicians prove to be incompetent time-servers and corrupt bunglers. They cannot compare with Nordenholt, the man of instant decisions, who plans to save a fraction of the British people by transferring his elect to a nitrogen-producing area in the Clyde Valley. Meanwhile, in a wicked world the worst begins to happen:

> The old civilisation went its way, healthy on the surface, full of life and vigour, apparently unshakeable in its power. Yet all the while, at the back of it there lurked in odd corners the brutal instincts, darting into view at times for a moment and then returning into the darkness which was their home. Suddenly came the Famine; and civilisation shook, grew weaker and lost its power over men. With that, all the evil passions were unleashed and set free to run abroad. Bolder and bolder they grew, till at last civilisation went down before them ...[6]

As the ravages of *Bacterium diazotans* begin to decline, the survivors under the autocratic direction of Nordenholt set about the building of their new society. The appearance of the scientist as saviour and superman is yet another indication of the loss of optimism that begins to affect utopian fiction after the First World War. After 1918 destruction seems to matter more than construction. Nordenholt speaks in a violent and absolutist manner never heard before in the ideal states of the future:

> Most of our old troubles have solved themselves, or will solve themselves in the course of the next few months. There's no idle class in the Nitrogen Area; money's only a convenient fiction and now they know it by experience; there's no Parliament, no gabble about Democracy, no laws that a man can't understand. I've made a clean sweep of most of the old system, and the rest will go down before we're done.[7]

A great change has come over Utopia. The old faith in humanity has given place to a belief in the powers of an exceptional individual, a saviour far above the rest of the community in determination and intelligence, who is the only conceivable means of creating the ideal state. This loss of faith led to the ultimate, despairing vision of that last day when the great cities are empty and all the works of man are in ruins.

The period between the two world wars was the time when the Last Man, or the last human group, inherited the earth. This symbolic act of annihilation was a judgment on the weakness and wickedness of mankind. It was a movement from nationality to humanity. Most of all, it was a collective repudiation of the doctrine of progress, since these disaster stories brought history to an abrupt end in the exemplary catastrophes so carefully described in *Der Pestkrieg*, *Der Bazillenkrieg*, *La Guerre microbienne*, *The Collapse of Homo Sapiens* and *The Last Man*. Year by year they passed on their brief message with remorseless regularity; and that message was a dramatic version of what Bertrand Russell—one amongst many—wrote in *Icarus, or the Future of Science* in 1924:

> Science has not given men more self-control, more kindliness, or more power of discounting their passions ... Men's

> collective passions are largely evil; far the strongest of them are hatred and rivalry directed towards other groups. Therefore at present all that gives men power to indulge their collective passions is bad. That is why science threatens to cause the destruction of our civilisation.[8]

That threat became the whole reality of *Gay Hunter* where the Scottish writer, J. L. Mitchell, showed how 'all the skill with machine and tool, the giant bridges and the great furnaces, the airships that clove the sky, the silent, gigantic laboratories ... ended with a naked savage with a flint and steel lighting a fire to cook his food amid the ruin of a desolate England.'[9]

The spontaneous and intuitive reactions of the disaster stories show how faithfully the tale of the future has played the part of *Doppelgänger* to industrial civilization. From the beginning of this literature in the eighteenth century, the imaginative projection of future events served as an intermediary between technological progress and the expectations of society. From the general experience of the seemingly beneficent progress of the applied sciences many apprentice sorcerers drew the pleasing conclusion that humanity could go forward with confidence into the future, for ever adapting to and for ever absorbing one change after another.

And then, after the unexpected catastrophes of the First World War had destroyed the golden link between progress and posterity, the tale of the future became the shadow theatre of the Western world. To the stereotype of the Last Man on Earth a succession of original writers added the other classic myths of the post-war period; they presented their grave new world with an even choice between disaster and despotism, between the victory of the robots in Capek's play, *R.U.R.*, and the triumph of totalitarianism in Zamyatin's *We*. The moral of *R.U.R.*, which was first staged at the National Theatre in Prague in 1921, is spoken by the General Manager of Rossum's Universal Robots. He foresees how the robots—that is, the indiscriminate application of technological invention—will bring about the end of human history: 'Mankind will never cope with the Robots, and will never have control over them. Mankind will be overwhelmed in the deluge of these dreadful living machines, will be their slave, will live at their mercy.'[10] The inhuman, immensely powerful and intelligent robots were created in the image of a mankind gone mad; for part of Capek's intention in the play was to look at 'the young scientist, untroubled by metaphysical ideas; scientific experiment is to him the road to industrial production ... Those who think to master the industry are themselves mastered by it. Robots must be produced although they are, or rather *because* they are, a war industry.'[11]

Whilst the Czech audiences watched the robots take over the world, Yevgeny Zamyatin was trying to find a publisher for his most seditious account of crime and punishment in The One State. Zamyatin had no luck with *We* in the Soviet Union; it was the end of the Civil War and the Red Army was then engaged in dealing with the sailors of the Kronstadt naval base who had asked for free Soviets and a constituent assembly. That clash between freedom and authority was the main theme of *We*; and it has been at the centre of all the dark prophecies of the tyranny-to-come that have been a staple of futuristic fiction during the last fifty years. The terror story came direct from the long Russian experience of autocracy. The suppression of freedom in the cause of human happiness had been foretold in *The Brothers Karamazov*, where Dostoievsky gave the Grand Inquisitor these prophetic lines about the necessary rigours of the despotic paradise:

> And they will have no secrets from us. We shall allow or forbid them to live with their wives or mistresses, to have or have not children – according to whether they have been obedient or disobedient – and they will submit to us gladly and cheerfully. The most painful secrets of their conscience, all, all they will bring to us, and we shall have an answer for all.[12]

Zamyatin was well able to confront the Grand Inquisitor. As a one-time Bolshevik, who had left the party before the Revolution, he could draw on a more than adequate knowledge of totalitarian theory; and as a translator and a most perceptive critic of H. G. Wells, he knew how to use the apparatus of the scientific world of the future in his description of The One

State. The obedient, uniformed citizens of the twenty-sixth century are trained from their earliest years to observe the Tables of Hourly Commandments and to keep holy the Day of Unanimity; the Guardians ensure that the nameless numbers fulfil their duties and on great occasions the Benefactor comes down from the clouds, 'the new Jehovah in an aero, just as wise and cruelly loving as the Jehovah of the ancients'.

The biblical parallels are deliberate. Zamyatin reveals a dark paradise in which unswerving conformity to the regulations is a perpetual victory over human nature; and like any paradise it is girt round with a wall, the Green Wall, which separates untamed nature from the total regularity of the mechanical utopia. 'Oh, the great divinely limiting wisdom of walls, of barriers', writes D-503. 'Man ceased to be a wild animal only when we had built the Green Wall, when we had isolated our perfect machine world from the irrational, hideous world of trees, birds, animals.'[13] Within the inner sanctuary of the mind they have tamed the wild; the state treats dreams and any other manifestations of selfhood as forms of mental illness. But there are rebels in Eden; and the hesitant Adam of a new dispensation, D-503, discovers unsuspected capacities for emotion in himself when the female I-330 tempts him with forbidden things. 'But the main thing,' she says, 'is that I feel perfectly safe with you. You are such a darling fellow–oh, I feel certain of that–that you won't even think of going to the Bureau and reporting that I, now, am drinking liqueur, that I am smoking.'[14] For such a woman D-503 feels that paradise would be well lost; and, like another and more famous fallen angel, D-503 chooses not to serve the almighty and everlasting state. In his mind he rejects the observances that knit together all the citizens in one communion and fellowship: 'It was clear to me that all were saved, but that there was now no salvation for me–*I did not want to be saved*.' By that act D-503 became the first of the modern rebels against the super-state, the precursor of the Savage in *Brave New World* and of Winston in *Nineteen Eighty-Four*. In the ecstatic freedom of an unpermitted sexual relationship D-503 discovers the essential differences between self and society, and he begins to perceive that personality cannot be the product of society.

The frequent biblical references and the tight patterning of *We* show that Zamyatin intended to outline a moral geometry through the episodic theorems that mark the progress of D-503 from initial self-awareness to acts of rebellion. Zamyatin had devised the first of those schematic representations in which the protagonists act out the clash between political power and personal freedom with the uncompromising opposition of Vice and Virtue in a morality play. It is authority versus the individual; it is the knowledge of good and evil that leads to an inescapable choice between the material benefits of the absolute state and the emotional needs of the human being. The issue emerges with total clarity when the Benefactor sends for D-503; and Zamyatin sees to it that the reader will get the point, since he portrays the Benefactor as 'a bald-headed, a Socratically bald-headed man' in blasphemous imitation of Lenin. The Benefactor speaks like the author of the New Economic Policy:

> I ask you: what have men, from their swaddling-clothes days, been praying for, dreaming about, tormenting themselves for? Why, to have someone tell them, once and for all, just what happiness is – and then weld them to this happiness with chains. Well, what else are we doing now if not that?[15]

There are to be neither martyrs nor heroes. For the sake of the lesson the state has to triumph, so that Zamyatin can leave the final decision, the ultimate choice, to the reader. The rebel conforms: 'The next day I, D-503, appeared before the Benefactor and imparted to Him all I knew about the enemies of our happiness.' The story then closes with D-503 watching unmoved the torture and death of the woman he had loved. The last sentence in the book is an ironic challenge to the reader – 'For rationality must conquer.'

Those last words have reverberated throughout the tale of the future ever since they first appeared in an English translation in 1924. Zamyatin had opened the great debate about the direction of technological society, and his theme proved so topical that it gave a subject to the new medium of the film. In 1926 the German film director Fritz Lang produced in *Metropolis* a series of brilliant visual images of oppression and revolt in a tyrannical society; and the message for the 1920s came across in the black-and-white capitals of the last caption: THERE CAN BE NO UNDERSTANDING BETWEEN THE HANDS AND THE BRAIN UNLESS THE HEART ACTS AS MEDIATOR.

More films followed – *À nous la liberté* by René Clair in 1934 and *Modern Times* with Charlie Chaplin in 1936; and these warning visions have continued their variations on the Zamyatin theme up to the present day. The script-writers have kept pace with, and drawn material from, the many parables, allegories, cautionary tales and dark prophecies that taught the whole world how to understand the crucial differences between the Savage and Mustapha Mond in *Brave New World*, between Winston and Big Brother in *Nineteen Eighty-Four*. The symbolic situations in this new admonitory fiction worked out a novel typology of human anxieties through which many able writers have looked into the problems of the individual in an epoch of constant technological innovation. It is indicative of the eminently social role of futuristic fiction and a measure of the many apprehensions of the last half-century that all these visions of the oppressive society abrogate the sacred covenant between progress and mankind. They are reflections of a cardinal idea that Nicholas Berdyaev first stated in *The End of Our Time* in 1923 and Aldous Huxley chose of set purpose as the epigraph to *Brave New World* in 1932: 'Utopias are more realizable than those "realist politics" that are only the carefully calculated policies of office-holders, and towards utopias we are moving. But it is possible that a new age is already beginning, in which cultured and intelligent people will dream of ways to avoid ideal states and to get back to a society that is less "perfect" and more free.'

That theory of 1923 was proved right; the progressive, technological utopia has been a rarity in futuristic fiction since the end of the First World War. Although H. G. Wells managed to produce yet another world state out of the cataclysms of the New Warfare in *The Shape of Things to Come*, most of the utopias rejected the old compact between science and society. A new college of prophets revived the arcadian survival myth of Hudson's *A Crystal Age* in their accounts of the simpler life that would follow on the obliteration of industrial civilization. The titles spoke for themselves – *Theodore Savage*, *Deluge*, *Dawn*,

Woman Alive, *The Empty World*, *Three Men make a World*—all collaborating within an ideal system of rewards and punishments that consigns *Homo technologicus* to the pains of the coming conflagration and consoles the frantic survivors with the experience of resurrection in a better world. Like the author of *Anthem*, Ayn Rand, they weep for mankind: 'Thus did all thought, all science, all wisdom perish on earth. Thus did men —men with nothing to offer save their great number—lose the steel towers and the flying ships and the power wires.' And there shall be a new earth after the survivors have learnt the lesson of the past: 'I shall take my food from the earth by the toil of my own hands. And the toil of my own hands will create my flowering domain in the wilderness. I shall learn many secrets from my books, and I shall find the rest. And slowly, through the years, I shall rebuild the wonders of the past.'[16]

The years between the two world wars was a time of renewed myth-making; it was the time when the tale of the future first introduced those premonitions of ecological disaster and world catastrophe that have nowadays become a commonplace of international journalism. Fearful wars, inhuman tyrannies, the collapse of society, the end of humanity—these were imaginative codes by which the mind could find intelligible patterns in a time of troubles. Thus, the tales of the coming tyranny found their model in the practices of Stalin and Hitler; and the disaster stories were apt illustrations of what the eminent scientist Sir Daniel Hall had to say about 'The Pace of Progress' in the Rede Lecture to the University of Cambridge in 1935: 'No longer is it a sound basis for Government to assume that life will be carried on in the near future as it has been in the immediate past. It has become a commonplace that the march of science is no longer wholly beneficial, but is developing aspects destructive of our accustomed economy.'[17]

That truism of modern futurology was an original idea in the 1930s; and it was the inspiration of the most remarkable prophecies of that decade—*Last and First Men*, *Last Men in London*, *Star Maker*. In these books Olaf Stapledon began a new phase in the development of futuristic fiction: he placed the Earthmen for the first time in the context of a galactic history and he introduced the myth of the perpetuation and constant improvement of the human race through culture after culture, from Venus to Neptune, as far as the last of the human species 2,000 million years hence. From the ruinous wars of the late twentieth century and the end of European power Stapledon followed the imagined sequence of great empires and new religions to the emergence of new kinds of human being. The intention was to create a work of reassurance, as Stapledon made clear in the preface to *Last and First Men*:

> To romance of the future may seem to be an indulgence in ungoverned speculation for the sake of the marvellous. Yet controlled imagination in this sphere can be a very valuable exercise for minds bewildered about the present and its potentialities. Today we should welcome, and even study, every serious attempt to envisage the future of our race; not merely in order to grasp the very diverse and often tragic possibilities that confront us, but also that we may familiarize ourselves with the certainty that many of our most cherished ideals would seem puerile to more developed minds. To romance of the far future, then, is to attempt to see the human race in its cosmic setting, and to mould our hearts to entertain new values ... We must achieve neither mere history, nor mere fiction, but myth. A true myth is one which, within the universe of a certain culture (living or dead), expresses richly, and often perhaps tragically, the highest admirations possible within that culture.[18]

This programme for a new mythology of human destiny was typical of the self-consciousness that had come into the discussion of the future since the beginning of the First World War. The slaughter of the trenches, the fall of ancient monarchies, the establishment of a powerful communist state, the March on Rome, the Slump, the rise of the Nazis – these were changes of such magnitude and they raised such serious questions about coming things that they shattered the pre-war expectation of an uncomplicated passage from present to future. The sudden loss of old, familiar certainties brought on a crisis of confidence that has not yet been resolved. The fears for the future began with the devastating revelation of technological power in the

Great War, and these fears became real terror when the mushroom cloud above Hiroshima gave warning of even worse things to come. And yet, as the wireless and the aeroplane, then jet aircraft and satellite television, carried on their work of drawing the nations ever more closely together, they gave reasons for thinking that the technologies of world communication could be the promise of better things to come. This is to say, that the promise of the sciences, or rather the understanding of technological power, has been the primary influence in deciding attitudes to the future since the days of Watt and Condorcet. At first, in a world made over to the machine and dominated by the engineer, it seemed a glorious thing that the great world would, in Tennyson's phrase, 'spin for ever down the ringing grooves of change'. And this belief became dogma in the idea of progress, in the optimistic histories of the nineteenth century, in the constructive utopias of the future, and in the many predictions that foretold the greater peace and the growing prosperity of the planet Earth.

After 1918, however, when the oceanic and outward-looking nations of the first industrial revolution discovered the fatal flaw in the Faustian civilization they had created, they suffered a most painful reversal of their expectations. They abandoned the doctrine of progress, and some of their historians found that the auguries presaged the End of Empire, the End of our Time, the End of Civilization. Gloomy prophets revealed long-forgotten rhythms of doom and destiny in the cycle of human existence; and the gloomiest of them all, Oswald Spengler, spent five years of concentrated study in preparing a worthy burial-place for the idea of progress in the two volumes and 1,000 pages of *The Decline of the West.* Like Capek and Zamyatin, Spengler denied the last sacrament of hope to a dying civilization; and like them his last words committed the glories of science and Western society to an inevitable dissolution:

> For us, however, whom a Destiny has placed in this Culture and at this moment of its development – the moment when money is celebrating its last victories, and the Caesarism that is to succeed approaches with quiet, firm step – our direction, willed and obligatory at once, is set for us within

> narrow limits, and on any other terms life is not worth the living. We have not the freedom to reach to this or to that, but the freedom to do the necessary or to do nothing. And a task that historic necessity has set *will* be accomplished with the individual or against him.
>
> *Ducunt Fata Volentem, nolentem trahunt.*[19]

But that was only half the story of the fateful marriage between technology and progress. Although the imagination could no longer tolerate the bright visions of prodigious machines and perfect cities, although the tale of the future had for the most part become a chronicle of condemnation and rejection, the methods of exploring the future continued to grow in subtlety and technique. They have advanced from the rapid expansion of predictive writing in the 1920s to the recent establishment of such futurological institutions as the Gesellschaft für Zukunftsfragen, L'Association Internationale Futuribles, The Committee on the Next Thirty Years, The Kiev Symposium on Technological Forecasting, The Committee on the Year 2000.

The work of these institutions is the most complex stage so far in a serious and rational investigation of the future that reaches back to the enterprise of a British publishing firm in the 1920s. Between 1924 and 1931 Kegan Paul, Trench and Trubner repeated on a much wider scale what the *Fortnightly Review* had done in commissioning *Anticipations* from H. G. Wells at the beginning of the century. They brought out a series of monographs in which scientists, philosophers, poets, sociologists, theologians, and novelists set down their conjectures about life in the coming decades. The eighty-six titles[20] ranged from the less than serious – Robert Graves, for example, on *Lars Porsena, or the Future of Swearing and Improper Language* – to most perceptive and influential forecasts. One of these was *Daedalus, or Science and the Future* in which J. B. S. Haldane predicted the ascendancy of biology amongst the sciences and gave Huxley some important ideas for *Brave New World* in his account of future laboratory techniques for making test-tube babies. Another of these forecasts, *Paris, or The Future of War* by Captain Liddell Hart, gave the German General Staff some ideas about armoured warfare that developed into the *Durchbruch* tactics of the Panzer divisions. These monographs were part of a general eruption of predictive writing throughout the industrial countries. Some forecasters continued the practices of the military writers in the days before 1914, telling their readers what to expect in such works as—Julius von Bernhardi, *Vom Krieg der Zukunft*, 1920; Colonel J. F. C. Fuller, *Tanks in Future Warfare*, 1921; General Golovine, *The Problem of the Pacific*, 1922; Lieutenant-Colonel Velpry, *L'Avenir des chars d'assaut*, 1923; Hauptmann Ritter, *Der Zukunftskrieg und seine Waffen*, 1924.

In the same decade there was an even greater production of more general forecasts that gave their estimate of the state of the world in the years ahead.[21] In 1927 Anton Lübke published an admirable and extensive forecast of technological developments in *Technik und Mensch im Jahre 2000*. In 1928 the British journalist Philip Gibbs produced *The Day After Tomorrow*, in which he went over the latest ideas about air travel, atomic energy, extra-sensory perception, and so on. In 1929 the American architect Hugh Ferriss looked into the future of town planning in *The Metropolis of Tomorrow*; and in 1930 there was an unmistakable sign that the subject of the future had become a matter of popular interest when that notable barrister and politician, the Earl of Birkenhead, made his contribution to the discussion of the future with his forecast of *The World in 2030 A.D.*[22]

Birkenhead drew heavily on his immediate predecessors in the art of prediction, repeating the more fashionable expectations of the 1920s. By the year 2030 he thought it possible that the whole question of human heredity and eugenics would 'be swallowed up by the prospect of ectogenetic birth. By this is meant the development of a child from a fertilised cell outside its mother's body—in a glass vessel filled with serum on a laboratory bench.' The consequences of the laboratory baby would be a brave new world of selective breeding and servile social categories:

> By regulating the choice of ectogenetic parent of the next generation, the Cabinet of the future could breed a nation of industrious dullards, or leaven the population with fifty

> thousand irresponsible, if gifted, mural painters ... If it were possible to breed a race of strong healthy creatures, swift and ductile in intricate drudgery, yet lacking ambition, what ruling class would resist the temptation?
>
> Many of the arguments brought against slavery would be powerless in such a case; for the ectogenetic slave of the future would not feel his bonds. Every impulse which makes slavery degrading and irksome to ordinary humanity would be removed from his mental equipment. His only happiness would be in his task; he would be the exact counterpart of the worker bee.[23]

After that Birkenhead had second thoughts, and in his next paragraphs he goes on to argue that 'production will become so cheap, and, barring political or international upheavals, wealth will accumulate to such an extent, that the ectogenetic Robot will never be needed.' Improved production methods will make the ten-hour week the norm throughout the workshops of the world; and synthetic foods will end the Malthusian fear of starving millions:

> Synthetic foods and the production of animal tissues *in vitro* will finally set at rest those timid minds which prophesy a day when the earth's resources will not feed her children. Though all the inhabitable surface of the globe were inconveniently crowded, the millions of mankind could still be fed to repletion by such means.
>
> This second revolution in food production will consummate the decay of agriculture, which can only survive as a rich man's hobby. A man born in the twenty-first century may, in his wealthy rejuvenation, boast that the bread he eats is made from wheat which grows in his own fields. Ploughing may even become a fashionable accomplishment, and pig-keeping a charming old-world fancy. Probably, however, the synthetic foods of the next century will be so much more easily digested and appetising than their present equivalents, that agriculture will survive only in historical romances.[24]

By the end of his first chapter Birkenhead had forecast a world of peace and plenty. In 2030 cheap atomic energy makes possible vast engineering projects; light-weight atomic engines drive vertical-take-off aircraft at high speeds; the harnessing of wind and tidal power adds to the energy resources of the world; and many advances in medical knowledge have wiped out all epidemic diseases. Colour television in every home brings the latest baseball matches to the Americans, and 'the M.C.C. selection committee, in conclave at Lord's, will be able to follow the fortunes of an English eleven through the days (or weeks) of an Australian test-match.'

The bland chapters continue the complacent story, as the author steers a course between science fiction and contemporary forecasts. His course is straight ahead, progressive and unfailingly optimistic. There may be confederations of states in the twenty-first century. Canada and the United States may possibly be one of these, and the South American nations are likely to form some kind of union. The chances of a united Europe, however, are very slight; but all the signs point to a thriving British Empire and to the continued domination of India:

> British rule in India will endure. By 2030, whatever means of self-government India has achieved, she will still remain a loyal and integral part of the British Empire. Many longing eyes are cast upon her. Russia especially would be gratified to see the patient work of Bolshevist agents crowned by a successful Communist outbreak on a large scale. This outbreak will not occur. The future of India presents vast difficulties and anxieties, but the devoted careers of both Englishmen and Indians, working side by side, will surmount them, and India will grow to be a bulwark of strength, and an example of prosperity to the whole empire.[25]

The rest of the book continues with the good news of a peaceful and most prosperous world: university education for all; poverty is unknown; clean cities and noiseless motor-cars make a paradise of the urban environment; and the Sahara has been flooded to make a playground for the Europeans. The last word from the year 2030 is that men are preparing to go to Mars:

> By 2030 the first preparations for the first attempt to reach Mars may perhaps be under consideration. The hardy individuals who form the personnel of the expedition will be sent forth in a machine propelled like a rocket; and equipped with a number of light masts which can be quickly extended, like fishing rods, from its nose. The purpose of these will be to break the impact with which, granted all possible skill and luck, the projectile would strike the surface of the planet.
>
> The great problem which such an expedition will face, however, is the possibility of missing Mars altogether; and, having escaped from the Earth's gravitational field, of wandering aimlessly through space unable to find a planet where they can hope for asylum.[26]

Birkenhead had nothing original to say in his forecasts. All the propositions in *The World in 2030 A.D.*, from the flooding of the Sahara to test-tube babies, had been circulating in science fiction stories and in predictive literature before he decided to set down his ideas of the future. His choice of subject, however, shows that the shape of things-to-come was attracting the interest of many readers on both sides of the Atlantic towards the end of the 1920s.

In the United States the first science fiction magazine had already found a growing popular market–first, with *Amazing Stories* in 1926, and then with *Science Wonder Stories* in 1929. Both magazines were the work of Hugo Gernsback, a Luxembourger by origin, who emigrated to the United States in 1904. This inventive and versatile man began by manufacturing storage batteries for the new automobile industry, and then in 1908 he founded *Modern Electrics*, the first wireless periodical in the history of publishing. In 1913 this became the *Electrical Experimenter*, and in 1920 the growing volume of science fiction stories caused Gernsback to call his magazine *Science and Invention*. By 1923 he thought there was a market for a magazine devoted entirely to what he called scientifiction, and in 1926 he tested his belief by publishing *Amazing Stories*, the first of many science fiction magazines that have since then carried on a flourishing business in meeting the popular demand for tales of the future. From the start they exploited the contemporary enthusiasm for novelty; and they did much to reinforce the general expectation of coming changes through their stories of the feelies, television, atomic energy, space travel, robots, and all the other anticipations that composed the popular image of the future in the 1930s. Today they serve to mark the acceleration in the rate of change that has in recent years led to many anxious books about the problems of adjusting to rapid social and technological developments. But in the 1930s, so Aldous Huxley thought, there was still plenty of time:

> In 1931, when *Brave New World* was being written, I was convinced that there was still plenty of time. The completely organized society, the scientific caste system, the abolition of free will by methodical conditioning, the servitude made acceptable by regular doses of chemically induced happiness, the orthodoxies drummed in by nightly courses of sleep-teaching–these things were coming all right, but not in my time, not even in the time of my grandchildren ...
>
> Twenty-seven years later, in this third quarter of the twentieth century A.D., and long before the end of the first century A.F., I feel a good deal less optimistic than I did when I was writing *Brave New World*. The prophecies made in 1931 are coming true much sooner than I thought they would. The nightmare of total organization, which I had situated in the seventh century after Ford, has emerged from the safe, remote future and is now awaiting us, just around the next corner.[27]

Huxley wrote *Brave New World* in the post-war style of futuristic fiction. The helicopters, music machines, stereoscopic feelies and the rest of the furniture in A.F. 632 came from the common stock of science fiction topics in the 1920s. In like manner the horrors of life in a regimented future society had been a subject for satirical prophecies since the last months of the First World War. The first of these were: Owen Gregory, *Meccania: the Super-state*, 1918; Rose Macaulay, *What Not*, 1919; Leslie Beresford, *The Great Image*, 1921. Moreover, the English translation of Zamyatin's *We* in 1924 and the publication of E. M. Forster's *The Machine Stops* in 1928 had provided valuable lessons in the handling of the dystopian projection. Like these

43–44 After the First World War the most frequent image of the future is the vision of disaster – the world conquered by the robots in Capek's *R.U.R.*, and humanity transformed into the obedient serfs of Fritz Lang's *Metropolis*.

45–46 As the technologies advance in the 1920s, they provide ideas for the voyage to the Moon in Fritz Lang's *Die Frau im Mond*, and for imaginative forecasts of transatlantic flying machines.

47 The swift development of air transport led to forecasts of the airships and passenger planes that would be sure to come.

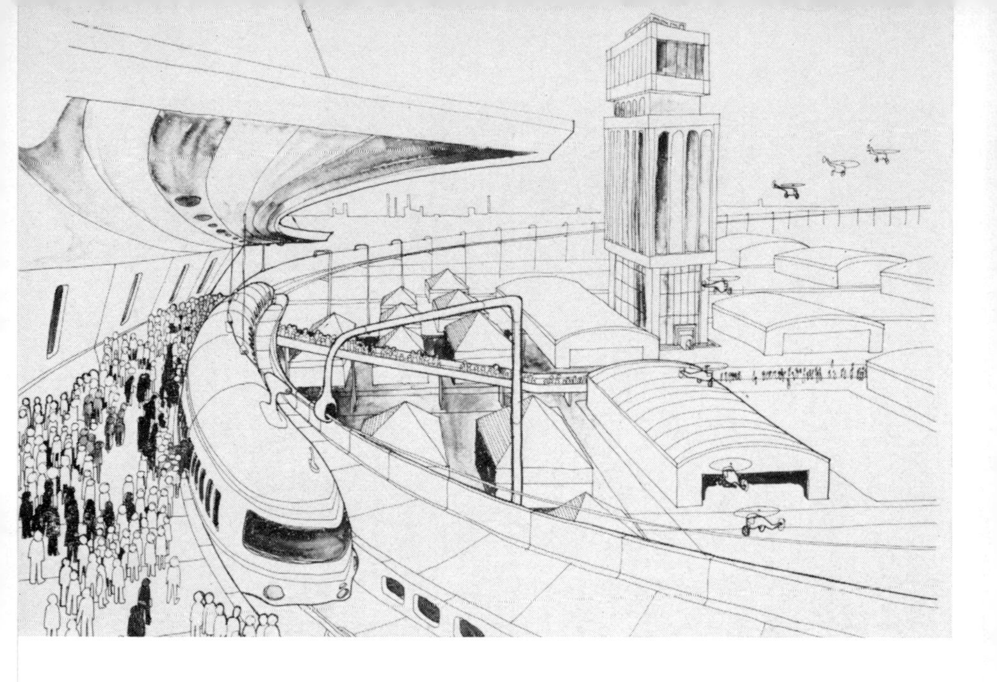

52–53 By the 1930s it was generally agreed that complex transport systems and vast skyscraper cities would be the norm for life in the future.

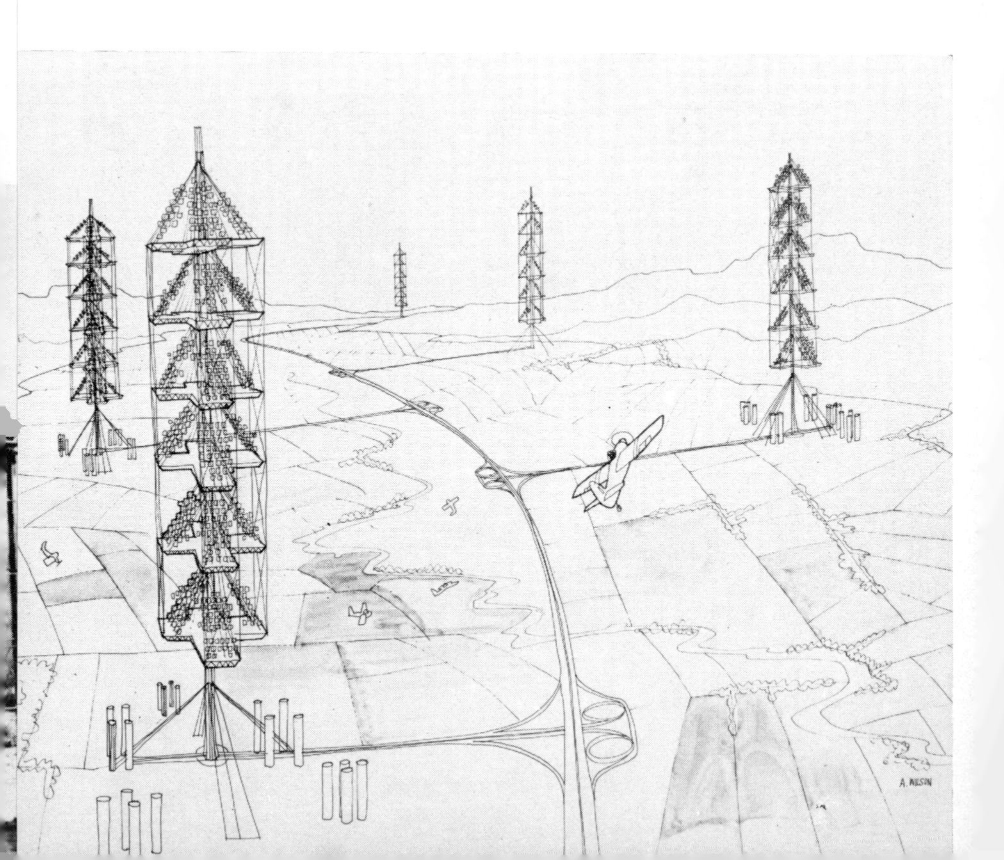

54–55 After the Second World War the classic statement about the future was the despair and decay of life in *Nineteen Eighty-Four*. Technology would be applied to the rigid control of all citizens.

56–57 The imagination ends where it began – with the anticipation of future warfare and with peaceful visions of technology in the service of mankind.

two writers Huxley had started from the Wellsian designs for the World State. Throughout *Brave New World* there is abundant evidence that Huxley had a thorough knowledge of Wells's utopian writings; and it is clear from Huxley's fiction and essays in the 1920s that he had long pondered the possibilities of the regulated world community.

In his first novel, *Crome Yellow* of 1921, Huxley introduced Mr Scogan as a means of exposing the positivist delight in technological progress and social regimentation. The resemblance between Scogan and Wells was deliberate. Like Wells, Scogan's 'nose was beaked, his dark eyes had the shining quickness of a robin's'; and Wells's high-pitched voice becomes the 'thin, fluty, and dry' conversation of Mr Scogan. He exults in the gifts of the applied sciences and he looks forward to a time in the course of the next few centuries when 'an impersonal generation will take the place of Nature's hideous system. In vast state incubators, rows upon rows of gravid bottles will supply the world with the population it requires.'[28] After that anticipation of *Brave New World* Scogan makes another entrance when he repeats the Wellsian doctrine that the men of intelligence can no longer leave the world to the direction of chance. Organization is the answer, and the Rational State is the means according to Mr Scogan:

> ... human beings will be separated out into distinct species, not according to the colour of their eyes or the shape of their skulls, but according to the qualities of their mind and temperament. Examining psychologists, trained to what would now seem an almost superhuman clairvoyance, will test each child that is born and assign it to its proper species. Duly labelled and docketed, the child will be given the education suitable to members of its species, and will be set, in adult life, to perform those functions which human beings in his variety are capable of performing.[29]

Ten years later Huxley was engaged on the details of the caste system in his brave new world; and in the September of 1931 he wrote that he had 'been very much preoccupied with a difficult piece of work—a Swiftian novel about the future, showing thc horrors of Utopia and the strange & appalling effects on feeling, "instincts" and general *Weltanschauung* of the application of psychological, physiological & mechanical knowledge to the fundamentals of human life'.

It is ironic that in *Brave New World* Huxley repeated the success of H. G. Wells in the 1890s; for Huxley wrote at a time when many earlier developments in futuristic fiction had made readers familiar with the themes of his book. It is even more ironic that *Brave New World*, which gained part of its strength from the evident reversal of Wellsian values, was a far better book than any of the Wellsian ideal states. Huxley had found an exciting and provocative way of expressing the anxieties and fears that ran through so many of the post-war tales of the future. He avoided the expository manner of utopian fiction and concentrated on an episodic style that moved the reader rapidly from one surprise to another. The constant shock of discovery is one way of making the reader draw the desired conclusion that things are not quite as wonderful as they seem in the world of Our Ford. Other methods which Huxley exploited with great effect were the motion picture techniques of cutting from one scene to another, of mixing dialogue, and of moving in on the major set pieces. In the famous opening paragraphs the narrative rolls forward like a moving camera on the film set of the Central London Hatchery. The opening sequence moves through the Fertilizing Room as the D.H.C. for Central London leads in the new students; then a visualization of the three hundred Fertilizers at their work as the text moves from the incubators across to the racks of test-tubes on their way to the Bottling Room; and then the description pans across the meticulous operations of the Bottle-liners and the Matriculators.

> In the Bottling Room all was harmonious bustle and ordered activity. Flaps of fresh sow's peritoneum ready cut to the proper size came shooting up in little lifts from the Organ Store in the sub-basement. Whizz and then, click! the lift-hatches flew open; the Bottle-Liner had only to reach out a hand, take the flap, insert, smooth-down, and before the lined bottle had had time to travel out of reach along the

> endless band, whizz, click! another flap of peritoneum had shot up from the depths ...

Brave New World is the classic story of futuristic fiction in the period between the wars. Huxley tells the world that man cannot live by technology alone, and he proves his point by the seeming inevitability of the social and mechanical projections that establish the context of life in the days of Our Ford. He conveys his message by the simple device of playing the Savage opposite Mustapha Mond. The Savage is the reader's idea of a natural, instinctive human being. He responds in an understandable way to the novel experiences in his strange new world. He wants love but not sex; he prefers the revelations of literature to the demands of social uniformity and political stability; he wants at all times to be himself. Thus, the operative field of Huxley's satire is the clash between the reader's consciousness of his own personal freedom and the artfully contrived dislike of the well-ordered, harmonious society of the future which cannot take account of an individual's desires and preferences. That issue runs through the story and at the end of the book it is the subject for a most famous debate between the Savage and Mustapha Mond. The Savage repeats the traditional values of Western civilization when he cries out: 'I want God, I want poetry, I want real danger, I want freedom, I want goodness. I want sin.' The World Controller asserts the necessities of the planned society, telling the Savage that his desires mean he is 'claiming the right to be unhappy ...'

> Not to mention the right to grow old and ugly and impotent; the right to have syphilis and cancer; the right to have too little to eat; the right to be lousy; the right to live in constant apprehension of what may happen to-morrow; the right to catch typhoid; the right to be tortured by unspeakable pains of every kind.

It was a year of portents when *Brave New World* came out in 1932. In Germany the Brüning government had been dismissed and Adolf Hitler was waiting on the events that would make him Chancellor. In the United States Roosevelt had been elected president on the promise of a 'New Deal' that would rescue the nation from the economic recession. And Oswald Spengler came in right on cue with his latest book, *Man and Technics*, in which he told his contemporaries that they were living through 'no mere crisis, but the *beginning of a catastrophe*'. All things organic, he said, are dying in the grip of organization. The artificial and mechanical has destroyed the natural and spontaneous. 'This machine-technics will end with the Faustian civilization and one day will lie in fragments, *forgotten*–our railways and steamships as dead as the Roman roads and the Chinese wall, our giant cities and skyscrapers in ruins like old Memphis and Babylon.'[30] The only way for industrial civilization was to endure the worst, 'to hold on to the lost position, without hope, without rescue, like that Roman soldier whose bones were found in front of a door in Pompeii'.

Ever since the end of the First World War many writers–American, British, Czech, French, German, Russian–had preached the one lesson that science had changed the conduct of life. Historians, forecasters, philosophers, and writers of fiction agreed that the world was moving towards a great crisis and that, in particular, another war could prove the final catastrophe. These fears were the origin of Spengler's prognostics and they provided material for the numerous accounts of the troubles-to-come that appeared in the fiction of the 1930s. One of the more original of these was *Public Faces* by Harold Nicolson, which was first published in the Faustian year of 1932. Nicolson had reflected on the contemporary anxieties about the powers of science and the dangers of technological warfare; and with an ingenuity that would have been creditable in H. G. Wells he based his story on the development of the atomic bomb.

When the British decide to test the bomb in the Atlantic, the United States cruiser *Omaha* is sunk by accident and a tidal wave causes the deaths of thousands in South Carolina. What follows is a sign of the great changes that had come over the tale of the future after 1918. Nicolson's story is not a political tract in favour of his own country; it is a polemical projection in support of all mankind. The text of the British declaration marks the extent of the change:

In connection with the disaster which has occurred in the Western Atlantic, the British Government make the following announcement:

(1) The wreck of the United States Scout Cruiser *Omaha*, as also the sinking of H.M.S. *Albatross* and the s.s. *Calamares*, and the tidal wave which broke upon the coast of South Carolina, were not, as contended by the United States Government, caused by experiments in any form of depth-charge.

These accidents were caused by the experimental release from Rocket Aeroplane I.A. of an atomic bomb.

(2) The British Government admit that the destructive range of this bomb had been seriously underestimated. It had been assumed that the atomic bomb would possess a range of destruction not wider than thirty miles in circumference. It is now clear that the range of destruction for each bomb is between seventy and eighty miles.

(3) On learning of the disaster the British Government at once issued orders that no rocket or other aeroplane shall, in time of peace, carry any atomic bombs: all existing stocks have been isolated in such a manner as to render impossible the recurrence of any further accidents.

(4) The British Government recognise, none the less, that full reparation is due to the United States for the damage which has regrettably been caused. They are prepared to submit the incident to any impartial tribunal and to abide by any reasonable award which such a tribunal may pronounce.

(5) The British Government, moreover, recognising that these new and potent engines of destruction are inimical to existing civilisation, are prepared (subject to the condition outlined below) to pledge themselves to destroy within a period of six months their existing stocks of atomic bombs and to manufacture no further bombs in the future.[31]

Since those prophetic words about the atomic bomb there have been no profound changes in the tale of the future. The principal anxieties of the 1930s have become the present-day visions of more fearful wars and the Orwellian terrors of the police state. The practice of forecasting, however, has seen a rapid development since the October of 1932 when Herbert Hoover received the report of the National Resources Committee to the President of the United States on *Recent Social Trends*. The Committee began by saying that 'in September 1929 the Chief Executive of the nation called upon members of this Committee to examine and to report upon recent social trends in the United States with a view to providing such a review as might supply a basis for the formulation of large national policies looking to the next phase in the nation's development. The summons was unique in our history.' More than that, it was unique in human history; it was the first large-scale attempt at a collective examination of the factors most likely to affect the future of a great nation. The third paragraph summed up the reasons that called the National Resources Committee into existence, and since then similar reasons have obliged almost all nations to establish their own committees for the investigation of the future:

> The first third of the twentieth century has been filled with epoch-making events and crowded with problems of great variety and complexity. The World War, the inflation and deflation of agriculture and business, our emergence as a creditor nation, the spectacular increase in efficiency and productivity and the tragic spread of unemployment and business distress, the experiment of prohibition, birth control, race riots, stoppage of immigration, women's suffrage, the struggles of the Progressive and the Farmer Labor parties, governmental corruption, crime and racketeering, the sprawl of great cities, the decadence of rural government, the birth of the League of Nations, the expansion of education, the rise and weakening of organized labor, the growth of spectacular fortunes, the advance of medical science, the emphasis on sports and recreation, the renewed interest in child welfare – these are a few of the many happenings which have marked one of the most eventful periods in our history.[32]

By 1932 forecasts of the future had become so frequent that the bibliographers began to list them under various classifications – *Anticipations*; *Aviation, future of*; *Forecasts, social, political, scientific*; *Future warfare*; *Utopias, see Forecasts*. A large number of these forecasts appeared in the United States, where a pragmatic people, accustomed by national tradition to the building of a future, took to the business of prediction with enthusiasm. The author of an American publication of 1939, *Science, Today and Tomorrow*, gave these reasons for the general interest in the future: 'My excuse for predicting, as well as attempting to elucidate, lies in the growing interest that the public displays in the social implications of science ... Indeed, a whole literature on what is called "the impact of science on society" has been produced within the last decade.' Within twelve months thousands of millions of human beings were to discover what the impact of science would be upon their societies.

The last reports from the tale of the future in 1939 were prophecies of destruction and desolation: Commandant Cazal, *L'Afrique en flammes*; P. G. Chadwick, *The Death Guard*; T. Craig, *Plague over London*; C. Marsh, *And Wars Shall Cease*; A. Marvell, *Three Men Make a World*; E. Ramseyer, *Airmen over the Suburb*; A. Schubert, *Weltenwende durch Gas*; D. Wheatley, *Sixty Days to Live*. Nevil Shute described the air raids of the coming war in *What Happened to the Corbetts*. In *The Hopkins Manuscript* R. C. Sherriff told how greed and folly destroyed Europe, as 'its last wretched inhabitants starved to death amid the ruins of their once noble cities'. And according to General Ared White in his *Attack on America* it was the beginning of the end for the United States after invading forces 'had landed at the mouth of the Columbia River on the Oregon coast, taken the antiquated coast forts from the rear with a few platoons of infantry, and were proceeding up the river toward Portland'. The last word, however, remained with H. G. Wells. In the first months of the war he told a friend that he was in poor health and was writing his epitaph. It would be quite short: 'Just this – God damn you all; I told you so.'

10

Science and Society, 1945–2001

The military operations of the Second World War—so unprecedented in scale, so devastating in effect—left the participants searching for words to explain the extraordinary transformations through which they had lived. So many certainties had vanished in those dangerous years when great armies, fleets and air forces were engaged in actions that thundered across the continents of a small planet. Tank battles in the Caucasus, desert warfare tactics in North Africa, the operation of fleet-trains in the Pacific, the running battle with the submarine in the North Atlantic, and the mushroom cloud above Hiroshima—these repudiated the old faith in the benefits of technology. The names of Belsen and Katyn, the inhuman schemes of Himmler and Beria, the devastation of Coventry and Dresden—these denied all past hopes for the progressive improvement of mankind.

The course of the Second World War multiplied the fears for the future. As the combatants grew accustomed to the regular introduction of new weapons, they realized that for the first time in the history of warfare an unseen struggle was going on in the Operational Research units and the laboratories of the belligerent nations. Everywhere scientists devoted themselves to supporting the fighting men with better fragmentation bombs, improved signals equipment, more advanced radar systems, faster and more manoeuvrable fighting vehicles, the invention of the flying bomb and jet aircraft. Later on, when the full story of the Manhattan Project and the V2 rockets began to be told in the immediate post-war years, it seemed self-evident that a most dangerous disruption had occurred between science and society. The conscript citizens, who had welcomed the applied sciences when they came bearing new means of warfare, turned into civilians and at once began to talk of a fatal flaw in the grand design of science. In the first decade after the war it seemed that the logic of technological development and the moral failures of society pointed to a dead-end for mankind.

There were many explanations. One of the first, and the saddest of them all, came from H. G. Wells. In the revised edition of *A Short History of the World*, which appeared one month after his death in 1946, Wells wrote a last testament for his idea of progress; and he punished mankind for not living up to his expectations by extinguishing all hope for the future of *Homo sapiens*:

> The seventieth chapter brought this History up to the year 1940. Since then a tremendous series of events has forced upon the intelligent observer the realization that the human story has already come to an end and that *Homo sapiens*, as he has been pleased to call himself, is in his present form played out. The stars in their courses have turned against him and he has to give place to some other animal better adapted to face the fate that closes in more and more swiftly upon mankind.[1]

One of Wells's younger contemporaries was C. E. M. Joad, of Birkbeck College in the University of London and the principal popularizer of philosophy in his day. In the second decade of the century he had embraced the doctrines of Shaw and Wells, but by 1949 the experience of two world wars compelled him to proclaim that human wickedness was the source of all evil in the world:

> As the melancholy history of the last thirty-five years unfolded itself, the liberal, optimistic view of history and the expectation of continual progress which it engendered, tolerable in the first decade of the century, came to seem increasingly untenable. A time came when I could hold it

> no longer. Indeed, in retrospect, one was left to wonder how it could ever have stood the test of the most cursory acquaintance with the facts of human history ... Finding myself in the end no longer able to write off man's 'sinfulness' – there is no other word for it – as a mere by-product of circumstances, I came to a conviction of the truth of the Christian doctrine of original sin, and to believe with the Ninth of the Thirty-Nine Articles that 'original sin' ... is the fault and corruption of every man, that naturally is engendered of the offspring of Adam.[2]

There were simpler explanations. Yet another of Wells's contemporaries, Somerset Maugham, ascribed the postwar pessimism to the direct consequence of moving out from the security of a more tranquil and more stable period in human history:

> The world has always been a place of turmoil. There have been short periods of peace and plenty, but they are exceptional, and because some of us have lived in such a period – the later years of the nineteenth century, the first decade of the twentieth – we have no right to look upon such a state as normal. Man is born into trouble as the sparks fly upward: that is normal, and we may just as well accept the fact.[3]

A very different explanation came from a young French woman who had followed the changing conditions of the war in constant anguish and in a most heroic unity of suffering with the afflicted. Simone Weil found that the triumphs of the applied sciences had obscured the proper orientation of human life; and in anticipation of the lesson Aldous Huxley was to teach in *Ape and Essence*, she wrote that material progress had created a false vision of the future:

> For several centuries we had lived upon the idea of progress. Today, suffering has almost eradicated that idea from our minds. So nothing now prevents us from seeing that it is not founded on reason. It was believed to be linked with the scientific conception of the world, though it is contradicted by science, as also by authentic philosophy. The latter

> follows Plato in teaching that the imperfect cannot give rise to the perfect or the less good to the better ... We cannot be made better except by the influence upon us of what is better than we are. What is better than we are cannot be found in the future. The future is empty and is filled by our imagination. Our imagination can only picture a perfection on our own scale. It is just as imperfect as we are; it does not surpass us by a single hair's breadth.[4]

The conclusions of H. G. Wells and Simone Weil mark the polarities in opinion and belief that provided the principal themes in the tale of the future through two decades, from *Das Glasperlenspiel* by Hermann Hesse in 1943 to Kurt Vonnegut's ironic treatment of human failings in *Cat's Cradle* of 1963. It was the period of the greatest originality in the writing of futuristic fiction; for this most multiform of genres, by returning to its origins in the hopes and fears of mankind, renewed those powers of immediate and compelling communication that have always characterized the various species of this prophetic literature. It was the beginning of the Atlantic era in the evolution of the genre. It was a time for all the talents, when the form of futuristic fiction attracted many gifted writers – Franz Werfel, Ernst Jünger, Aldous Huxley, Robert Graves, Anthony Burgess, Ray Bradbury, Pierre Boulle, Evelyn Waugh, William Golding, Angus Wilson, John Updike, Kurt Vonnegut. These employed the tale of the future as the most effective means of commenting on the situation of their time. They adopted a prophetic, admonitory or revelatory style so that they could analyse or anathematize the weaknesses of twentieth-century society; and some of them – Kurt Vonnegut, Anthony Burgess, and Aldous Huxley in particular – seemed bent on creating a new social homiletics about the deceits and delusions of modern man.

The German contributions to the new literature of the future were amongst the most vigorous and original in the history of the genre. Out of the terror and devastation of the Third Reich there came uncompromising affirmations of human fallibility and a compulsion to assert the primacy of the spiritual. The style of the new fiction was a formal verdict against those expectations of human perfection and continuing material improvement that had called the image of the constructive world state into being in the days of Edward Bellamy, Theodor Hertzka and H. G. Wells. In the profound parables of Hermann Hesse in *Das Glasperlenspiel* and of Franz Werfel in the *Stern der Ungeborenen* the time-scale shifts from the near future of the Victorian utopias to far-off epochs when the people of very different societies have no accurate recollections of the twentieth century. Distance lends a desired severity to the view of modern man; and the great gaps in time – a thousand years for Hermann Hesse and thousands of years for Franz Werfel – are ways of pointing to the gulf between the real and the ideal. Thus, the journey to the remote future becomes an odyssey of the spirit in a time of troubles. These writers turn to the future because there is no conceivable locus in the contemporary world for all they have to say about the workings of good and evil in human society.

The meditative calm of the narrative in *Das Glasperlenspiel* works as counterpoint in the rhythmical unfolding of a rare harmony; for this *Meisterwerk* of Hermann Hesse, as Thomas Mann called it, has a good claim to be considered the most imaginative and subtle parable in the entire history of futuristic fiction. It is a prose poem in celebration of creative activity; it is a penetrating study of the individual and the community, of the sacred and the profane, of intellect and nature. The first of these major themes appears in the third paragraph where Hesse writes that the disciplined thinkers of Castalia consider personality 'is something quite different from what the biographers and historians of earlier times meant by it ... '

> For them, and especially for the writers of those days who had a distinct taste for biography, the essence of a personality seems to have been deviance, abnormality, uniqueness, in fact all too often the pathological. We moderns, on the other hand, do not even speak of major personalities until we encounter men who have gone beyond all original and idiosyncratic qualities to achieve the greatest possible integration into the generality, the greatest possible service to the suprapersonal. If we look closely into the matter we shall see

> that the ancients had already perceived this ideal. The figure of the Sage or Perfect One among the ancient Chinese, for example, or the ideal of Socratic ethics, can scarcely be distinguished from our present ideal; and many a great organization, such as the Roman Church in the eras of its greatest power, has recognized similar principles. Indeed many of its greatest figures, such as St. Thomas Aquinas, appear to us – like early Greek sculptures – more the classical representatives of types than individuals.[5]

This is a new language and a new attitude of mind in the tale of the future; it is a looking backward to origins in order to go forward to the reformed way of life in the land of Castalia. That mountainous region is set aside from the rest of the world thanks to the influence of the Castalian Order which emerged from the final epoch of warfare. The special function of the Order is to provide educators for the world; its greatest strength derives from the cultivation of the Glass Bead Game, which is a metaphor for the promotion of cultural and spiritual harmonies. These represent a perfect and benevolent totalitarianism:

> For our distinction is to cherish the true sanctuary of Castalia, its unique mystery and symbol, the Glass Bead Game. Castalia rears pre-eminent musicians and art historians, philologists, mathematicians, and other scholars. Every Castalian institute and every Castalian should hold to only two goals and ideals: to attain the utmost command of his subject, and to keep himself and his subject vital and flexible by forever recognizing its ties with all other disciplines and by maintaining amicable relations with all. This second ideal, the conception of the inner unity of all man's cultural efforts, the idea of universality, has found perfect expression in our illustrious Game.[6]

The Game is a complex system of signs and grammatical forms that draws on the sciences and the arts; it is essentially 'a mode of playing with the total contents and values of our culture'. The Game forms the intelligence and heightens the awareness of the players. The great weakness, however, is that it encourages in the Castalian Order a tendency 'toward smugness and self-praise, toward the cultivation and elaboration of intellectual specialism'. Intellect, scholarship and aesthetic refinement, Hesse suggests, can never make a satisfactory whole of the individual. For personal fulfilment and for the salvation of the social order it is necessary to transcend the self, to develop 'that limited freedom of decision and action which is the human prerogative and which makes world history the history of mankind'. Translated into bald English, Hesse seems to be engaged in unashamed sermonizing; but the incantatory quality of his language and the powerful significations of the symbolism raise his concern for contemporary society to the level of an all-embracing and compassionate understanding. The dramatic tension in the story comes from the life and deeds of Joseph Knecht who, as his name suggests, becomes the servant of humanity in his supreme office as Master of the Game. He comes to realize that abstraction and aestheticism are not enough, that Castalia belongs to the world of human actions and political decisions. So, Knecht ends by calling on his Order to maintain the truth, to surrender their privileges, and to dedicate themselves in a time of troubles to the education of the young. That is 'where the basis for the cultural life of the country is to be found, not in the seminars or in the Glass Bead Game ... More and more we must recognize the humble, highly responsible service to the secular schools as the chief and most honourable part of our mission.'

Human destiny, the condition of mankind and the perils that threatened the Western tradition – these were the main concerns of the best post-war tales of the future in German. In 1945, a few weeks before he died, Franz Werfel completed the last chapters of his admonitory vision of the remote future in the *Stern der Ungeborenen*. In the year 101,945 his time-traveller finds a functional society in which the politicians, workers, churchmen and scholars have their designated roles; he observes an uprising of the jungle people who are deserters from the world society; and he comes to the conclusion that his 'once vague belief in original sin, in man's fall through disobedience, has been profoundly strengthened' as a result of his journey to the future.

This schematic representation of twentieth-century experience is typical of German futuristic fiction after the war. The tale of the future becomes a hospital of the mind. Allegories of human wickedness, parables of social morality, surrealistic representations of recent history, hopeful and despairing prophecies of other times and other societies – these pour out in an extraordinary unburdening of the mind. The common theme of such outstanding writers as Stefan Andres, Ernst Jünger, Walter Jens, Hans Erich Nossack and Arno Schmidt is that the State has become Leviathan; the unnatural monster devours its own children. They write with anguish of the world they have seen and of a society in which man is regarded as an object, as a functional unit in a remorseless political system. Walter Jens, for example, follows in the tradition of Zamyatin and Kafka in his *Nein. Die Welt der Angeklagten* of 1950. In fact, he enlarges on the ideas of Kafka in *The Trial* and *The Castle* by describing a time when society is divided into the three classes of the Plaintiffs, the Witnesses and the Judges. In this hateful state, in which the Palace of Justice is the centre of terror, all the citizens begin in the class of the Plaintiffs. By denouncing their friends they can advance to the ranks of the Witnesses, and from these they may be selected for the highest office of Judge. The parallels are both pointed and painful.

The general strategy is to expose the failings of the twentieth century by singling out the abuses of power – either of man over his fellows or of mankind over nature. So, the Rhinelander Stefan Andres transposes the catastrophes of modern times to the symbolic history in his admirable trilogy of *Die Sintflut*. He opens in *Das Tier aus der Tiefe* of 1949 with the political agitator who becomes a dictator; he goes on to the outbreak of a destructive war, and he ends the trilogy with *Der Regenbogen* of 1952, when the flood of evil and devastation sweeps away the despotic society. What Stefan Andres says about political man has an evident parallel in what Arno Schmidt chose to say about natural man in *Die Gelehrtenrepublik* of 1957. There is the same concentration on the gravest faults of contemporary civilization, the same desire to find the most imaginative and forceful means of moral instruction. The story is the choreography of a powerful detestation which moves rapidly from one symbolic situation to another, as Schmidt shows how an outraged nature may punish the follies of undisciplined humanity. The narrator, an American journalist, opens the drama of self-discovery in the year 2008 after an atomic war has devastated the world. Most of Europe and parts of North America are Hominid Areas. There the mutants live – the Centaurs, the poisonous Scorpion men and the Butterfly Men with human heads. The narrator travels through fabulous lands to an artificial island, a floating Parnassus, where all the artists and scientists worth saving have been preserved; and the narrator's experiences during the perilous journey allow Arno Schmidt to make some telling points about love and hatred, man and nature, and the lunacies of the Cold War.

The quality of literary achievement in the best examples of the postwar *Zukunftsroman* is impressive. It is evident that many writers had felt themselves impelled to work out a comprehensive geography of the imagination so that they could explore the dark continents of contemporary experience. The main lines of this development appear most clearly in the work of Ernst Jünger who first made a reputation for himself with his war diary, *In Stahlgewittern*, in 1920. There he recorded the course of the First World War seen through the eyes of a soldier who had been decorated with the rare *Pour-le-Mérite* in recognition of heroic service in the trenches. His books in the 1920s were in praise of German militarism and during the 1930s he wrote in favour of the totalitarian state. After the Second World War, however, Jünger went through a change of ideas and began to proclaim the Christian virtues of love and forgiveness; and in his prophetic utopia, *Heliopolis* of 1949, he sought to sum up his thoughts on ethics and politics. Once again a German writer had chosen the far-off future as the setting for an allegory of modern times that carries a considerable weight of philosophizing and theologizing. In the city of the sun all power is divided equally between the aristocratic Prokonsul (that is, Hindenburg or the reality of power through the control of the armed forces) and the plebeian Landvogt, who is a Hitler-figure representing the will and political force of the people. Both sides wait in readiness for the inescapable conflict between them; and the anti-hero of the story travels through the land reporting on the various answers to the great questions of morality and power as these are revealed in the apparatus of the future society.

The undertones of Zamyatin, Spengler, Huxley and Orwell in these German tales of the future speak of a postwar community of thought in European fiction. The contemporaneous developments in American and English fiction repeat on a more extensive scale and in a more vehement manner the recurrent preoccupations of the anti-utopian literature of the 1920s and 1930s. Although the Americans had never suffered under a totalitarian government, and had never known the horrors of concentration camps, mass executions, starvation and the appalling destructiveness of air bombardments, they agreed with the Europeans that in a hateful age things were going from damnable to desperate. In 1949 Huxley and Orwell produced the first major postwar myths of the destroyed society in *Ape and Essence* and the despotic society in *Nineteen Eighty-Four*. And the Americans followed in 1952 with Kurt Vonnegut's ironic version of a computer-controlled United States in *Player Piano*, and then in 1954 there was Ray Bradbury's classic invention of the book-burning firemen in *Fahrenheit 451*. In those days, when men looked for a sign, they could choose from Big Brother, the Arch-Vicar of Belial, the Epicac XIV National Computer, and that cruel promise of perpetual servitude in the shape of the eight-legged infallible questing beast, the Mechanical Hound, 'leaping out like a moth in the raw light, finding, holding its victim, inserting the needle and going back to its kennel to die as if a switch had been turned'.

These first ominous visions were symptomatic of a great fear, of a mood of pessimism that had been growing since the 1920s and had been exacerbated to the point of despair by the experiences of the Second World War. For the Europeans these admonitory projections were a continuation and intensification of those intellectual and moral anxieties that had first revealed themselves in the prophetic nightmares of Zamyatin and Capek, the gloomy prognostications of Oswald Spengler, and the journey through the Wasteland in the poetry of T. S. Eliot. The measure of this advance in fear and trembling appears in the harsh severity that develops between the relatively humane management of society in *Brave New World* and the degraded

Belial worshippers in the post-catastrophe society of *Ape and Essence*. The stages of this advance in a dreadful self-awareness appear in the poetry of Wystan Auden who saw the Second World War as a visitation on all for the moral, social and political failings of the past. In the *New Year Letter* of 1941 he wrote about his time in verse that anticipates the thoughts of Orwell and Huxley:

> The future that confronts us has
> No likeness to that age when, as
> Rome's hugger-mugger unity
> Was slowly knocked to pieces by
> The unco-ordinated blows
> Of artless and barbaric foes,
> The stressed and rhyming measures rose:
> The cities we abandon fall
> To nothing primitive at all
> This lust in action to destroy
> Is not the pure instinctive joy
> Of Animals, but the refined
> Creation of machines and mind;
> As out of Europe comes a Voice
> Compelling all to make their choice,
> A theologian who denies
> What more than twenty centuries
> Of Europe have assumed to be
> The basis of civility,
> Our evil Daimon to express
> In all its ugly nakedness
> What none before dared say aloud,
> The metaphysics of the Crowd,
> The Immanent Imperative
> By which the lost and injured live
> In mechanized societies
> Where natural intuition dies,
> The international result
> Of Industry's Quicunque Vult,
> The hitherto-unconscious creed
> Of little men who half succeed.[7]

By the end of the war Auden had arrived at the point when, in marked contrast to his earlier poetry in the 1930s, he had begun to assert the paramount values of Christian belief. In his Christmas Oratorio of 1945, *For the Time Being*, he used the events and doctrines of the Annunciation, Incarnation and Nativity for a series of meditations on human society and human destiny. He spoke directly and with conviction about the prospects of mankind in this vale of tears:

> We know very well we are not unlucky but evil,
> That the dream of a Perfect State or No state at all,
> To which we fly for refuge, is a part of our punishment.
> Let us therefore be contrite but without anxiety,
> For Powers and Times are not gods but mortal gifts from God;
> Let us acknowledge our defeats but without despair,
> For all societies and epochs are transient details,
> Transmitting an everlasting opportunity,
> That the Kingdom of Heaven may come, not in our Present
> And not in our Future, but in the Fullness of Time.
> Let us pray.[8]

Auden had touched on the principal concerns that provided the subject matter of the more influential and widely read tales of the future on both sides of the Atlantic during the first two decades after the war. These were the powers of technology, the possibilities of despotism, the potential conflict between the individual and society, and the moral or political ideas that could decide the future of the world. A unity of mind never known before in this literature caused many able writers to create horrifying visions of the police states of the future, the urbanized hells of mass societies, the atomic wars that destroy civilization and the universal cataclysms that wipe out the human race. It was a time of little hope and great fear. Everyone had read the book of recent history, and all could make the deduction that out of great evils there might be worse to come. As Aldous Huxley said with complete truth of Orwell's *Nineteen Eighty-Four*, it was 'a magnified projection into the future of a present that contained Stalinism and an immediate past that had witnessed the flowering of Nazism'.[9]

It seemed that history was on the side of the pessimists, for there were some historians who followed Spengler in foretelling the end of Western civilization. According to the American scholar Pitirim Sorokin, 'the twilight of the Sensate phase of Western culture' had begun. The shades of night were closing in: 'Before our very eyes this culture is committing suicide. If it does not die in our lifetime, it can hardly recover from the exhaustion of its creative forces and from the wounds of self-destruction.'[10] In 1941 Sorokin published the last of the four large volumes of his *Social and Cultural Dynamics*, and he closed his survey with his last words on the future – thirteen predictions that foretold the social, political, cultural, aesthetic and moral disintegration of Western civilization. However, out of the agonizing process of *crisis–catharsis–charisma* would come the sunburst of the purified idealistic Culture of the future, and the evolutionary sequence would go forward to even greater achievements.

A more comprehensive and a far more spectacular prophecy came from the English historian Arnold Toynbee who completed twenty-seven years of work in 1954 when he published the last four volumes of his monumental *Study of History*. A more copious writer than Spengler or Sorokin, Toynbee felt he had been called to sit in judgment on the course of human history. He pursued his thesis on the rise and fall of the great societies through ten volumes, 6,000 years of history and an investigation of twenty-one civilizations, from the arrested and abortive to the defunct, developing, and disintegrating. In the 1940s it required some effort of mind to resist the moral vigour, the charming discursive style and the immense learning which Toynbee brought to his account of the inevitable decline of great civilizations. The grand dramatic theme of the rise and fall of nations was congenial to the pessimism of the postwar period. When Toynbee wrote of the ebb and flow of civilizations, which rose or fell in time with the mechanisms of Challenge-and-Response, he seemed to speak directly to the condition of the times. All who read Toynbee knew that they were living through what he called a Time of Troubles. The end of the European empires, the emergence of new nations in Asia and Africa, the division of the world between the Russians and the

Americans, the hard recovery of Europe, the great social transformations going on throughout the industrial countries, the cruel anxieties of the Cold War – all these immense changes raised the gravest questions about the future of Europe and, indeed, about the future of the world.

Among the signs of those times are the singular consistency of argument that marks the prophetic school of history and the even greater coherence of imagination in the contemporary tales of the future. For the historians the apocalyptic succession starts from that day in 1911 when Oswald Spengler first became aware of 'a historic change of phase' in the forward movement of European history. The aim of Oswald Spengler, Walter Schubart, Pitirim Sorokin, and Arnold Toynbee was to arrive at a final synthesis of human history. By this means they hoped to create a synoptic device for the understanding of past, present and future. Their predecessors in the writing of histories – Macaulay and Buckle, for example – had explained the development of civilization as the onward rush of the great wave of progress. This was 'the liberal, optimistic view of history', which C. E. M. Joad embraced in his youth only to abandon in middle age. In the complex argumentation of Spengler and Toynbee the rejection of that simple dogma was the recognition of a crucial difference between the unquestionable fact of material progress and the evident failure to achieve any significant form of moral progress. Toynbee, for instance, claimed 'that it could not fail to be evident to a Western historian, taking an observation in A.D. 1952, that progress in Technology ... '

> ... so far from being a guarantee of progress in virtue and happiness, was a challenge to it. Each time that Man increased the potency of his material tools, he was increasing the gravity of the moral consequences of his acts and was thereby raising the minimum standard of the goodness required of him if his growing power was not to turn to his destruction.[11]

The theories of Sorokin and Toynbee exorcized the ghosts of nineteenth-century historicism. In the name of their own historical determinism they anathematized those ancestral preconceptions that had only a century before so happily prophesied an eternity of progress for the European nations. 'Yesterday Hitler marched into Prague,' wrote the poet Edwin Muir. 'The nineteenth century sowed the whirlwind that we are reaping ... '

> If I look back over the last hundred years it seems to me that we have lost more than we have gained, that what we have lost was valuable, and that what we have gained is trifling, for what we have lost was old and what we have gained is merely new. The world might have settled down into a passable Utopia by now if it had not been for 'progress'.[12]

This lamentation for things lost has been the central preoccupation of the prophetic histories and of the more purposeful tales of the future ever since the outbreak of the First World War. The writings of Spengler, Sorokin and Toynbee represent a movement throughout Western countries to comprehend the condition of the twentieth century; and the parallel developments in dystopian fiction manifest a comparable effort of the imagination both in Europe and the United States to explicate the moral confusions of modern times. These frightening worlds of the future compose a unique body of prophetic parables, instantly recognizable and accepted everywhere as hard currency in the exchange of ideas about freedom and the citizen, science and society, morality and politics. Thus, the philosophizing tendencies of Spengler, Sorokin and Toynbee have their counterpart in the fantasizing techniques of Arno Schmidt, Aldous Huxley and Kurt Vonnegut. The imaginative writers seek to codify tendencies in their societies within a fictional vocabulary that speaks of despotism and disaster. Through elaborate typologies – Belial and the Bomb, Big Brother and O'Brien – and through most eloquent symbols of oppression and catastrophe – Newspeak and the Wastelands – they create an environment of the imagination in which it is possible to measure the worst tendencies and gravest failings of the day against ideas of personal integrity and common humanity.

As a general rule these minatory stories begin in the near future, near enough to frighten the reader with the terrors of tomorrow. The most favoured style of narrative advances through the realistic description of future conditions to the high point of a dramatic confrontation in which the protagonists argue out the rival necessities of citizen and state, or contest the issues between technological gadgetry and moral goodness. Thus, Aldous Huxley opened the debate in the February of 1949 with a sermon on the wickedness of political man and the profligate misuse of scientific knowledge in *Ape and Essence*. The story opens in the blasted deserts of southern California after the atomic bombs of the Third World War have wiped out most of mankind. In the year 2108 the members of the New Zealand Re-Discovery Expedition to North America find 'that what was once the world's largest oasis is now its greatest agglomeration of ruins in a waste-land. Nothing moves in the streets. Dunes of sand have drifted across the concrete.' In the exemplary manner of the preacher Huxley shows how men in their folly have made a hell of their world; and in one of the cruellest visions of recent fiction he reveals that the degraded inhabitants of the Californian desert have made a god in their own image. They worship Belial, the Lord of the Flies; and his Arch-Vicar speaks directly for Friar Huxley when he explains how the Great Blowfly has worked his will in human history:

> Take the First World War, for example. If the people and the politicians hadn't been possessed, they'd have listened to Benedict XV or Lord Lansdowne – they'd have come to terms, they'd have negotiated a peace without victory. But they couldn't, they couldn't. It was impossible for them to act in their own self-interest. They had to do what the Belial in them dictated – and the Belial in them wanted the Communist Revolution, wanted the Fascist reaction to that revolution, wanted Mussolini and Hitler and the Politburo, wanted famine, inflation and depression; wanted armaments as a cure for unemployment; wanted the persecution of the Jews and the Kulaks; wanted the Nazis and the Communists to divide Poland and then go to war with one another. Yes, and He wanted the wholesale revival of slavery in its most brutal form. He wanted forced migrations and mass pauperization. He wanted concentration camps and gas chambers

> and cremation ovens. He wanted saturation bombing (what a deliciously juicy phrase!); He wanted the destruction overnight of a century's accumulation of wealth and all the potentialities of future prosperity, decency, freedom and culture.[13]

After that commentary from the true *advocatus diaboli* there followed in the May of 1949 the unsparing Orwellian homilies of *Nineteen Eighty-Four*. Some of the worst experiences of the war years provided the inspiration for the projections in Orwell's story: the party apparatus, the perpetual war, the endless propaganda, the police patrols. His most original invention of *Newspeak* is the product of a logical imagination working on the nature of language and the enormities of totalitarianism. The vocabulary of *goodthink*, *crimethink*, *thinkpol*, *prolefeed* lacks the charm of novelty; it is too close to the verbiage of the Gestapo and the K.G.B. for comfort:

> The purpose of Newspeak was not only to provide a medium of expression for the world-view and mental habits proper to the devotees of Ingsoc, but to make all other modes of thought impossible. It was intended that when Newspeak had been adopted once and for all and Oldspeak forgotten, a heretical thought – that is, a thought diverging from the principles of Ingsoc – should be literally unthinkable, at least so far as thought is dependent on words. Its vocabulary was so constricted as to give exact and often very subtle expression to every meaning that a Party member could properly wish to express, while excluding all other meanings and also the possibility of arriving at them by indirect methods.[14]

The immediate commercial success of *Nineteen Eighty-Four* was a testimony to one of the most resourceful and rigorous imaginations to work in the medium of futuristic fiction since the best days of H. G. Wells. The furious controversies that followed between Stalinists and Trotskyites, between Socialists and Liberals, were proof that the Orwellian projection could be read as a verification of contemporary political tendencies. The rapid translation into foreign languages, private circulation in the Soviet Union and the immediate adoption of those powerful slogans about freedom and slavery – these were signs that Orwell had spoken to readers everywhere about the political and moral circumstances of the individual in the last years of Stalin. In the darkest period of the Cold War, when the Berlin Blockade and the war in Korea appeared as a confrontation between good and evil, the future world of Big Brother and Miniluv was often taken for a scenario in support of NATO. Despite the overt relationship with totalitarian systems, the story of *Nineteen Eighty-Four* is concerned more with the person than with the Party. The issue between Winston and O'Brien is the clash between despotic authority and the right of the individual to be himself, the right to arrive at his own understanding of the purpose of life. So, Winston takes the first fatal step of committing his private thoughts to a diary. 'It was not by making yourself heard,' Winston thinks, 'but by staying sane that you carried on the human heritage.' He went back to the table, dipped his pen, and wrote:

> *To the future or to the past, to a time when thought is free, when men are different from one another and do not live alone – to a time when truth exists and what is done cannot be undone:*
> *From the age of uniformity, from the age of solitude, from the age of Big Brother, from the age of doublethink – greetings!*[15]

From the primal sin of a private act Winston advances like a tragic hero to the inevitable doom of discovery and condemnation; and with Winston go the cardinal virtues of liberal democracy. In the interrogation rooms of the Ministry of Love he learns the whole gospel of the absolute insignificance of the individual. O'Brien in his office as grand inquisitor tells Winston that power for the sake of power is the end of the Party; and Winston has to realize 'that power is collective. The individual only has power in so far as he ceases to be an individual.' The absolute demands of absolute power are stated in language reminiscent of Dostoievsky and Zamyatin:

> Do you begin to see, then, what kind of world we are creating? It is the exact opposite of the stupid hedonistic Utopias that the old reformers imagined. A world of fear and treachery and torment, a world of trampling and being

> trampled upon, a world which will grow not less but *more* merciless as it refines itself. Progress in our world will be a progress towards more pain. The old civilizations claimed that they were founded on love and justice. Ours is founded upon hatred. In our world there will be no emotions except fear, rage, triumph, and self-abasement. Everything else we shall destroy. Already we are breaking down the habits of thought which have survived from before the Revolution. We have cut the links between child and parent, and between man and man, and between man and woman. No one dares trust a wife or a child or a friend any longer. But in the future there will be no wives and no friends. Children will be taken from their mothers at birth, as one takes eggs from a hen. The sex instinct will be eradicated.[16]

The anxieties and uncertainties of the 1940s have their monument in the austere and unrelenting prophecies of Orwell and Huxley. They probe the wounds of a civilization thought to be in the last stages of life. They keep their eyes resolutely on the gravest failings of the postwar period; they refuse to think of palliatives, of the promise of happier times to come or the fanciful dream of life in a golden past. They put the issue to their readers as an inescapable choice between anticipated terrors and adamantine principles. From the Sinai of their imagined future they hand down a law for the living. Mankind has to endure the worst, they imply, because there cannot be any going back along the road of urbanization and technological progress. The one hope for the future lies in self-transcendence: to deny the Great Blowfly within the human heart, to develop the heroic constancy that Winston never knew.

The Orwellian tale of terror belongs to the work of reconsideration that has been going on these last sixty years. The apocalyptic visions of the future have distinct affinities with the vaticinations of the prophetic historians. All of them search the past for clues to contemporary dilemmas. 'The situation of our time,' says Wystan Auden, 'Surrounds us like a baffling crime.' And then, as if called by magic, Robert Graves appears in the September of that same augural year of 1949; and in *Seven Days in New Crete*, that happy tale of wizardry and world peace, he presents his own findings on the human situation. Like Huxley and Orwell he begins in the common style of great wars and imminent disasters. His time-traveller discovers that the civilization of New Crete had emerged from the ferocious wars of the twentieth century:

> After a series of revolutions and minor wars, the close of the late Christian epoch was marked by an unusually savage struggle between the so-called Roman Bloc, consisting of the communist and semi-communist states of Western Europe and North America, and the so-called Orthodox Bloc consisting of neo-communist Eastern Europe and the Far East, both Blocs being nominally Christian. This war which, as usual, lasted far longer than had been expected, laid most of Western Europe waste ... [17]

By an ingenious adaptation of time-travelling conventions the narrator is summoned to the world of New Crete by the spells of a witch; and by an exquisite variation on past practice in futuristic fiction Robert Graves takes the utopian writers at their own estimation. He creates an ideal state out of the best that is known and thought in Western literature. The history of the great change begins from the time when an Israeli Sophocrat wrote a book, *A Critique of Utopias*, that greatly impressed his colleagues in Southern Europe, America and Africa ...

> From a detailed and learned analysis of some seventy *Utopias*, including Plato's *Timaeus* and *Republic*, Bacon's *New Atlantis*, Campanella's *Civitas Solis*, Fenelon's *Voyage en Solente*, Cabet's *Voyage en Icarie*, Lytton's *Coming Race*, Morris's *News from Nowhere*, Butler's *Erewhon*, Huxley's *Brave New World*, and various works of the twenty-first to the twenty-fourth centuries, he traced the course of man's increasing discontent with civilization as it developed and came to a practical conclusion: that 'we must retrace our steps, or perish'.[18]

The desperate survivors of the third millennium go back to the beginning. They establish a number of experimental communities in the areas that had escaped the devastation: 'These enclaves were to represent successive stages of the development of civilization, from a Palaeolithic enclave in Libya to a Late Iron Age in the Pyrenees; and were to be sealed off from the rest of the world for three generations, though kept under continuous observation by field-workers directly responsible to the Anthropological Council.' The Cretan community works out the answer to the problems of the good society, and within five hundred years their system has spread throughout the habitable world: no machines, no printed books, ritual battles with padded quarter-staffs, the worship of the Mother-Goddess, the cultivation of witchcraft, and the annual death of the Sacred King whose holy blood sanctifies the fields and ensures the peace of all. 'It's because of the awful holiness of this sacrifice that the New Cretan custom forbids the violent taking of life on any other occasion, even in war.'

For most of the narrative the reader is led to believe that this frolic of the future represents an ideal way of life, although the time-traveller at the outset notes that the Cretans 'lacked the quality that we prize as character: the look of indomitability which comes from dire experiences nobly faced and overcome ... Not only did they lack character, which the conditions of their life had not allowed them to develop, they lacked humour – the pinch of snuff that routs the charging bull.' In the last pages the prophet reveals his purpose: the Cretans are too secure, and they have to suffer hardship and sorrow before they can grow to the full stature of human beings. The time-traveller tells the Cretans that the Goddess has called him from the past in order to infect the too-perfect future. He comes to them as 'a seed of trouble'; his mission is to bring them 'a harvest of trouble, since true love and wisdom spring only from calamity'. And he returns to the happiness of true love in the twentieth century, crying out: 'Blow, North wind, blow! Blow away security.'

In the remote past before the First World War, in the fiction of Bellamy and Hertzka, the belief was that final happiness lay within the grasp of humanity. In the era of Hitler and Stalin, however, the lesson of the centuries seemed to be that fortitude is the better part of wisdom. The severe morality of the apocalyptic prophecy works its purpose after the manner of the events in *King Lear*. The authors open with the follies and discords that have torn society apart; and they end, as Shakespeare ends in the last act of his play, with summary executions, stern judgments and an indomitable resolution. Huxley, Orwell and Graves tell the world that there cannot be any going back. Indeed, the intention behind *Seven Days in New Crete* is to reject the proposition that 'we must retrace our steps, or perish'. And Graves ends with an affirmation of the calamitous condition of human life that recalls the last words of Edgar to the blinded Gloucester: 'Men must endure their going hence, even as their coming hither. Ripeness is all.'

In this way the pliant and responsive tale of the future assumed the shape and the burden of contemporary anxieties. For some twenty years, from the end of the 1940s to the late 1960s, the genre concerned itself with dystopian visions of despotism and disaster, and with ruthless repudiations of twentieth-century civilization–parables in which a nobler species replaces *Homo sapiens*, myths of rebirth in which the survivors rebuild the good society. The many similarities between authors are not evidence of borrowings. They mark the narrowness of the seams in which they choose to work and they point to an international consensus on the way in which the world seemed to be going in the worst years of the Cold War. Thus, the Orwellian device of *Newspeak* was the invention of a writer who had long considered the totalitarian methods of manipulating language in order to control the mind of the citizen. In fact, Orwell's fabrication of *The B Vocabulary* ('words which had been deliberately constructed for political purposes') was the first alarming message in an early-warning system that was then beginning to defend the necessary truthfulness of human communications. Orwell had anticipated some of the methods in the lurid futuristic fantasies of *The Naked Lunch* and *Nova Express*, where William Burroughs breaks down linguistic expectations in order to save language from the manipulators. '*Peoples of the earth, you have all been poisoned*' is the slogan of an American who rebels against the corrupt practices of the communications industry. 'Who monopolized Life Time and Fortune?' Burroughs asks with direct reference to the bland narcotism of the expensive magazines. 'Who took from you what is yours? Now they will give it all back? Did they ever give anything away for nothing?' The same points are central to the argument of Marshall McLuhan in *The Mechanical Bride* of 1951. He enlarges on the Orwellian proposition that in the world of 1984 words will be employed 'to impose a desirable mental pattern upon the persons using them'. He argues that the twentieth century 'is the first age in which many thousands of the best-trained individual minds have made it a full-time business to get inside the collective public mind. To get inside in order to manipulate, exploit, control is the object now.'[19]

At first sight the *Nadsat* vocabulary, which Anthony Burgess employs with such vigorous effect in *A Clockwork Orange* of 1962, might seem a derivative of *Newspeak*, whereas it belongs to the tradition of lexical inventions that goes back to More's *Utopia*. Orwell uses *Newspeak* to instance the depravities of the totalitarian mentality, but Burgess used *Nadsat* to distance the reader from the depravities of Alex and his companions. In his artful way Burgess puts the words of *banda*, *bitva*, *bugatty* and the rest into the mouths of the vicious young in order to set them apart by their social dialect from the Standard English, 'this very beautiful real educated goloss,' of their preceptors. By this means they are seen to act outside the standard morality of their society. Thus, Orwell, Burroughs and McLuhan represent the practical and political awareness of the circumscribing and conditioning uses of language that is a major interest of our time. And in like manner, Burgess, Tolkien and C. S. Lewis show a similar interest in the workings of language in so far as it operates as a code to carry the culture and modify the behaviour of any society. The vocabulary of C. S. Lewis in his space fiction and the invented languages in *The Lord of the Rings* trilogy, like the *Nadsat* phraseology, are the work of expert philologists. Indeed, the contrast between the imaginary languages of Tolkien and Burgess is a striking example of the differences between linguistic advantage and linguistic disadvantage. The *Nadsat* speakers are confined by their circumstances to a restricted and predictable code that cannot find any words to express concepts, whereas the nobler tongues of Middle-earth exhibit all the subtlety and flexibility of an elaborated code that is a suitable vehicle for rational communication. Although Burgess has little in common with Orwell, and there is literally a world of difference separating Burroughs from Tolkien and C. S. Lewis, all of them use their linguistic devices to illuminate matters of immense importance. They want their readers to see with a new insight that the word is the beginning and the end of civilized life. They say with Jean-Paul Sartre that the function of a writer 'is to call a spade a spade. If words are sick, it is up to us to cure them.'[20]

It is important to note that the Atlantic marks an instructive divide in the defence of civilized values. The English tale of the future has produced the most convincing prophecies of political oppression, and the Americans have written more widely and with greater effectiveness about the repressive potentialities of the mechanized society. These appear as nightmare fantasies of the computers and the robots that nullify the American dream of the free, spontaneous and heroic life. In a world dominated by machines there can be no place for the Minute Man, unless he is prepared to be a martyr. The impulse in this fiction—both technological and ideational—is profoundly American; for writers like Ray Bradbury, Kurt Vonnegut, John Barth, and Bernard Wolfe focus their stories on the ways in which technological developments can destroy prized relationships in society. One reason is that ever since the construction of the first commercial computer UNIVAC in 1950 the Americans have been living through the most extensive applications of computers in world history. They have in consequence engaged in the most serious debates about their uses and abuses, because the possibility of an automated world controlled by computers and data banks offends all that is personal and independent in the American tradition.

The analogue of Big Brother in American futuristic fiction is the all-powerful computer. It represents the worst elements in a malign system of total control which has to be destroyed, or from which the individual has to escape in order to discover a lost personality. So, the computer becomes the object of the greatest detestation—EPICAC XIV in Kurt Vonnegut's *Player Piano*, EMSIAC in Bernard Wolfe's *Limbo '90*, WESCAC in John Barth's *Giles Goat-Boy*. The authors project contemporary anxieties into their visions of an America that is no longer the sweet land of liberty. The title of Vonnegut's story states the casc of them all against the dangers of an industrial and political system in which the desire for progress and prosperity could lead to a regimented society of obedient citizens operating in witless harmony like the tinkling notes of a pianola. In the near future of *Player Piano* democracy and freedom seem to flourish as vigorously as ever, until the satirist reveals that the price of national prosperity is an I.Q. rating of 140. The future belongs to the experts, a small group. The rest are the unemployed who can volunteer for an easy time in an army that has nothing to defend or they can choose the comforts of a life with the Reconstruction and Reclamation Corps that has no identifiable task. That situation, transposed to the circumstances of World War III, provides the germ of Bernard Wolfe's *Limbo '90*; and in the postscript to his prophecy of 1990 Wolfe speaks for Vonnegut when he records that he is writing 'about the overtone and undertone of *now*–in the guise of 1990 because it would take decades for a year like 1950 to be milked of its implications'. And he speaks for all the anxious postwar prophets when he says that 'on the spurious map of the future ... I have to inscribe, as did the medieval cartographers over all the terrifying areas outside their ken: HERE LIVE LIONS.'

The lions attack mankind in World War III, which 'is the first homicidal chess game in which the full gaming board has been used and all the pawns thrown into action with perfect mathematical precision ... '

> It was bound to happen, of course. Once men stopped manufacturing goods, they began to manufacture machines. Whence EMSIAC, the god-in-the-machine, the god-machine.
>
> We could have predicted it. If we'd had our eyes open, we would have seen it coming. EMSIAC is simply the end development of something that's been threatening for a long time in human affairs, especially in modern times. Hobbes called it the Leviathan–I'd call it the Steamroller. War–this present war, this epitome of warness–is only the Steamroller come of age ... The flattening of the human spirit, I mean. What the steamroller does to the human spirit is immeasurably worse than anything shrapnel and atomic blast could possibly do to the human flesh, and infinitely more lasting.[21]

Wolfe describes the greedy, conscienceless exploitation of human beings as 'the smothering of the *I* by the *It*', a fighting phrase that summons every American to stand up like a second George Washington against the automated monarchies of technology and big business. The call resounds through the American science fiction of the 1950s in many stories about the mechanized worlds of the future where the American hero struggles like a latterday Laocoon with the menaces of the advertisement men, the communications media, robots and androids, omniscient computers and various cybernetic horrors. The more notable of these stories are: *The Humanoids*, in which Jack Williamson looks at the benign stagnation of life that the robots bring to a civilization; *The Space Merchants*, in which Frederik Pohl and C. M. Kornbluth narrate a famous story of Madison Avenue advertising in an overcrowded world; *Solar Lottery*, where Philip K. Dick turns the banal television quiz-game into a savage battle of winner-take-all; and in *The Caves of Steel* that most imaginative writer Isaac Asimov describes the time when all mankind has crowded into 800 vast cities, each of them with an average population of 10 millions.

> Each City became a semiautonomous unit, economically all but self-sufficient. It could roof itself in, gird itself about, burrow itself under. It became a steel cave, a tremendous, self-contained cave of steel and concrete.
>
> It could lay itself out scientifically. At the centre was the enormous complex of administrative offices. In careful orientation to one another and to the whole were the large residential Sections connected and interlaced by the expressway and the local way. Towards the outskirts were the factories, the hydroponic plants, the yeast-culture vats, the power plants. Through all the melee were the water pipes and sewage ducts, schools, prisons and shops, power lines and communication beams. There was no doubt about it: the City was the culmination of man's mastery over the environment.[22]

The more the world changes the more it grows worse – that is the principal lesson of recent futuristic fiction. Nowadays the account of life in the future speaks to the whole family of man in a common language of apprehensions and anticipations. The universal symbols of the Swastika, the Red Star, the mushroom cloud, the Sputnik and the Lunar Excursion Module have their fictional counterparts in the universally recognizable images of Big Brother, the desolate cities, and the galactic empires of a new interplanetary mythology. The appetite for this literature is matched only by the inventiveness of the largest number of the ablest writers ever to tell the tale of delight or disaster. The scale of production is evident in Frederik Pohl's claim that, although *The Space Merchants* 'probably isn't the most successful novel ever written ... with somewhere over ten million copies in print in about forty languages, it hasn't done really badly either'.[23] And there are the annual reprints of *Nineteen Eighty-Four* and the world audience for Stanley Kubrick's prodigious epic of *2001: A Space Odyssey* and for the picture of the damned in his film version of *A Clockwork Orange* by Anthony Burgess.

These visions of the future are signs of their times. They are a proof collective that technology decides the condition of life throughout the planet Earth and that change is the master of their fictional universe. All the varieties of futuristic fiction, from the gloomy dystopias to the complacent contemplation of *Homo cosmonauticus* triumphant throughout the galaxies, all carry on the unbroken confabulation between writers and readers that has developed during two centuries into a universally accepted mode for the appraisal of human possibilities. The direction of this development has been from great hopes to greater fears – from the chauvinistic dream of a worldwide *pax britannica* presented by a Tory propagandist in *The Reign of George VI* in 1763 to the nightmare images of universal destruction in that famous Kubrick film of 1963, *Dr. Strangelove; or, How I Learned to Stop Worrying and Love the Bomb*.

During the 1950s the tale of the future surrendered to the overwhelming terrors of 'the age of the assassins' which the French poet Rimbaud had foretold long ago. 'Never has *homo faber* better understood', said Sartre, 'that he has *made* history and never has he felt so powerless before history.'[24] And so, during the 1950s, in keeping with the movement of contemporary literature, the tale of the future finally abandoned the old faith in the power of intellect and in the progressive improvement of civilized life. What remains is a powerful desire to cry havoc at the dangers of a run-away world and the determination to pass on the news that salvation is still within the moral reach of mankind. Running through all the tales of the bad times that will come is a profound distrust in, often a contempt for, the abilities and the wisdom of political man. The mood is more urgent and the moral has a greater consequence than at any earlier period in the evolution of futuristic fiction. The principal strategy is to entrap the readers, to enclose them in an iron totalitarianism, to engulf them in the miseries of universal destruction; and the favoured point of departure is a headlong leap into affliction.

The profusion of invention is without parallel in the history of the genre; and most of the classic statements on the hateful future appear in the United Kingdom. It seems that the descendants of Charles Darwin and Herbert Spencer, no longer able to accept the idea of progress, have chosen to discharge their feelings of anger and anxiety by working through the entire paradigm of social and political parables. In 1952 there was J. W. Wall's excellent tale of 'the hundred and second year of the First German Millennium', *The Sound of his Horn*, when the Nazis have divided Europe between the ruling Aryans and – a familiar theme – the servile masses they can hunt like animals. In 1953 Evelyn Waugh brought out his ironic story of the decadent Welfare State, *Love among the Ruins*, where he writes with cruel wit about the seediness of life in the days when the over-zealous state holds no man responsible for his actions. And then in 1954 William Golding went back to the origins of human failings, and in *The Lord of the Flies* he took his title from the Beelzebub of *Ape and Essence* as a signal that he had chosen to join in the debate about the nature of man and the condition of society. Another sign of this conscious intervention is in the way Golding places the story in the near future after a nuclear war has broken out. The sins of the parents, as Golding shows, are repeated in the sinfulness of the children. The tropical island, where the aeroplane has left the children is the world of unspoilt nature which the children turn into a hell of their own making.

The flood of these prophetic stories went on without any slackening throughout the 1950s and into the late 1960s. For some authors the tale of the future was an opportunity for allegory or a convenient means of looking into varieties of behaviour rather than a serious pursuit of the possible. For example, L. P. Hartley chose for the subject of his *Facial Justice* of 1960 the deliberate suppression of all personal differences in a totalitarian Britain after World War III. Again, in 1961 Angus Wilson turned from the lively studies of middle-class attitudes that have made him famous, and he used the improbable idea of a war between Europe and the United Kingdom as the setting for a study of private loyalties and ferocious animosities in *The Old Men at the Zoo*. This was very different from the intention and the methods of Michael Young who gave a word to the language in *The Rise of the Meritocracy, 1870–2033*, that witty and plausible tale of rational solutions for the production problems of an industrial nation. His sources were the developments in secondary and university education in the 1950s, recent practice in intelligence and aptitude testing, and the fact that a high I.Q. is a national asset in a period of industrial innovation.

When the British people find that their productivity is declining – a familiar story – they decide that the remedy lies in finding the right person for the right occupation. The grand strategy turns on the theory that ability generates merit in any well-organized society, and that national success will follow when society sees to it that the citizens with the highest I.Q.s receive the best education and that the most lucrative posts go to the most efficient managers. The most rational schemes, however, can have undesirable results; and the unfailing moral rigour of prophetic literature begins to take over as the story reveals that by 1990 the nation has been divided into competent, highly rewarded meritocrats and the less competent masses. There are rumblings of protest, and history takes a new direction when 'shaggy young girls from Newnham and Somerville, instead of taking the jobs as surgeons and scientists for which their education fitted them, scattered to Salford and Newcastle to become factory workers, ticket collectors, and air hostesses'. For the sake of fraternal feelings and a nobler theory of social justice Michael Young, the President of the Consumers Association, rejects the best schemes of the consumer society.

One of the worst facts of contemporary society – the violence of the young – gave Anthony Burgess the central idea for a most remarkable tale of the future. The theme of *A Clockwork Orange* is the choice between good and evil in a time when the law encourages violence against the criminal in the form of brain-washing techniques. The narrator is the brutal, depraved but intelligent adolescent Alex who chooses evil for its own sake. His crimes find him out, and to cure him of his vicious habits he is sent for a course of aversion-therapy in the State Institute for Reclamation of Criminal Types. Which is the greater crime – the compulsory shock treatment and revulsion methods of Ludovico's Technique or the violent acts of a young criminal? The Burgess answer is to ask the question: 'Is the man who chooses the bad perhaps in some way better than a man who has the good imposed upon him?' The final answer comes from the character who has suffered most at the hands of the young ruffian:

> You've sinned, I suppose, but your punishment has been out of all proportion. They have turned you into something other than a human being. You have no power of choice any longer. You are committed to socially acceptable acts, a little machine capable only of good. And I see that clearly – that business about the marginal conditionings. Music and the sexual act, literature, and art, all must be a source now not of pleasure but of pain.[25]

In this way the extraordinary violence of our time is contained and considered within the moral dimensions of futuristic fiction. By projecting tendencies and possibilities into an imaginary future a writer is able to direct attention to the great war of man against man, of nation against nation. By the imaginative process of realizing in a convincing manner the peculiar circumstances of the future society a writer creates vivid images and memorable scenes that represent moral judgments on our life and times. Indeed, the tale of the future has been the conscience and the confessional of the Atlantic community since the 1940s. There can be no doubt that penitential works like

Nineteen Eighty-Four, *Die Gelehrtenrepublik*, *Limbo '90* and *A Clockwork Orange* castigate the deadly sins of the late twentieth century – lust, envy, fratricidal hatred, the extravagant exploitation of nature, the brutal suppression of human liberty, the reckless development of the most lethal weapons. The authors work with a missionary zeal through all the devices of fiction, from eloquent allegories of human weakness to realistic narratives of the last days of mankind. They toil through the desperate category of anxieties, afflictions and terrors in order to persuade the reader that society must reform – must find new ways of living – before it is too late.

For many authors *too late* meant the moment after the intercontinental missiles blast away from the Minuteman sites. The last and worst moment before the nuclear war was a major topic in futuristic fiction, and in the 1960s it appeared on television and film. In the United States there were many stories of the final war, some of them gaining instant notoriety as *Fail Safe* did in 1962. These advanced through the frequently rehearsed stages of the coming disaster from *Red Alert* to *After Doomsday*, *Alas! Babylon* and *The Long Tomorrow*. In the United Kingdom the same fears ran through the same sequence from *Extinction Bomber* and *Point of No Return* to the catastrophes described in *On the Beach*, *And So Ends the World*, *By Then Mankind Ceased to Exist*. The contemporary stereotype in this fiction begins when a surviving member of a government tells a nation of the indescribable disasters that have struck the country; and it ends with fearful scenes of the kind that appear in the television film *The War Game* which the B.B.C. banned in 1966.

It is an ironic commentary on modern times that one invention – television – should prove so terrifying a means of envisaging those other developments of recent years – total war and nuclear weapons. For this reason the directors of the B.B.C. felt it necessary to prevent the transmission of the film they had commissioned Peter Watkins to make. Their explanation was that:

> When the film was completed and screened for senior programme staff of the B.B.C. most of those who saw it

> were very deeply affected, and believed that it had the power to produce unpredictable emotions and moral difficulties whose resolution called for balance and judgement of the highest order. The horror of the film was, in their view, of an entirely different quality to that which is contained in the recognisably fictional presentation of some television films ... this horrifying effect is inherent in the pictorial presentation, rather than in the script itself. It was quite different in its degree of horror from anything which they had previously experienced.[26]

The documentary techniques of *The War Game* represent in an episodic and pictographic manner what the graphs and extrapolations had predicted in contemporary books on the subject of *War in the Atomic Age*, *Atomic Weapons in the Next War* and *On Thermonuclear War*. As the camera tracks across scenes of future devastation and moves in on the crumpled bodies and livid faces of the victims, the commentator remains silent. The spectators require no telling that they are watching the end of their world. The scenes act out what President Kennedy implied in his Inauguration Address on January 20th, 1961, when he spoke of the perils of living 'in an age where the instruments of war have far outpaced the instruments of peace'.

> The Polar DEW has just warned that
> A nuclear rocket strike of
> At least one thousand megatons
> Has been launched by the enemy
> Directly at our major cities.
> This announcement will take
> Two and a quarter minutes to make.
> You therefore have a further
> Eight and a quarter minutes
> To comply with the shelter
> Requirements published in the Civil
> Defence Code–section Atomic Attack.[27]

The jargon of overkill and megaton annihilation triggers the imagination into an instantaneous action replay. The mind races through the spools of past expectations, back from the

Cuban missile crisis of 1962 and the testing of the first hydrogen bombs in 1954, back to the outbreak of the Crimean War in 1854. In that year, when the world still rejoiced in the happy marriage between science and society, the mild and amiable writer Thomas De Quincey turned out an essay 'On War'. He foretold that 'the same cause which makes war continually rarer will tend to make each separate war shorter. There will, therefore, in the coming generations be less of war; and what there is will ... be indefinitely humanised and refined.' *Circumspice, si monumentum requiris.*

11

Into the Third Millennium

Today the futurologists of the world unite in preparing their scenarios for the tomorrow of the third millennium. Yesterday the historian J. B. Bury wrote an epilogue to the expectations of the first industrial age in the last pages of a famous book on *The Idea of Progress*. Looking at the world of 1920, Bury says that the great deceiver of mankind is the illusion of finality, the conviction that some theory or philosophy will be valid for all time. The idea of progress shares in the illusion. A stubborn belief in the continuity of human affairs makes it difficult to accept that 'the order of things with which we are familiar has so little stability that our actual descendants may be born into a world as different from ours as ours is from that of the pleistocene age'. The rigorous historian of the idea of progress then closes his book by prophesying that 'a new idea will usurp its place as the directing idea of humanity ... In other words, does not Progress itself suggest that its value as a doctrine is only relative, corresponding to a certain not very advanced stage of civilization?'

The record of futuristic writing confirms the prophecy. Since 1945 the revision of the doctrine has been particularly far-reaching in the unique codex of secular scriptures that reason and the new technologies called into existence in the second half of the eighteenth century. The most striking changes have been the explosion of dystopian fiction, discussed in the last chapter, and the metamorphosis of the ideal state which has gone through the contractions of a painful rebirth. Megalopolis has yielded place to utopia. The ideal life can only be reached imaginatively in the here-and-now of small communities or in the vastly changed circumstances of some remote future period. This symbolic rejection of the continuities of technological progress marks a movement from the material and political to moral and social considerations. It is in no sense a self-indulgent arcadianism. On the contrary, it is central to a rethinking of the place and purpose of mankind that goes from the behavioural engineering of the American psychologist B. F. Skinner in *Walden Two* in 1948 to the spiritualization of the evolutionary process in the writings of the Jesuit palaeontologist Teilhard de Chardin.

Since Buchenwald and Hiroshima the differences between society as it is and the society desired by the idealists have appeared so enormous that the small volume of utopian fiction has established a new literature of total discontinuity. This rejects the animosities and antagonisms of world politics for the cultivation of close personal relationships and loyalty to the chosen community. 'We have no imperialist policy—no designs on the possessions of others,' says the founder of the harmonious community in *Walden Two*. 'What is Walden Two but a grand experiment in the structure of a peaceful world? Point to any internationalist who really *knows* what sort of society or culture or government will make for peace. He *doesn't* know! He's only guessing!'[1]

Although there is a question mark above the conditioning techniques employed in *Walden Two*, the utopia deals with the theme of self-creation that appears in *Providence Island* by Jacquetta Hawkes, *Island* by Aldous Huxley and *The Dispossessed* by Ursula Le Guin. These represent a statement of personal values in the face of collective indifference which is also characteristic of contemporary existentialist philosophy. For example, Jacquetta Hawkes locates her ideal community on an undiscovered island to the north of New Guinea. The exceptional psychic powers of the Magdalenian inhabitants, all happily preserved from the curse of mechanical invention, confront the academic disciplines of anthropology and archaeology. The investigators learn their lesson. The islanders are not moral primitives. They are all that the twentieth-century citizen is not: tranquil in behaviour, happy in themselves, untroubled in their sexuality, well adjusted to their communal life, capable of expressing themselves in an admirable art and beautiful rituals.

The intention in the utopian fiction of the post-war years is to redeem the moral failures of our time by revealing a society in which the individual is helped to lead a life that is good in itself and can become better. This was undoubtedly Huxley's intention when he began the planning of *Island* in 1956. His scheme was to write 'about an imaginary society, whose purpose is to get its members to realize their highest potentialities. I shall place the fable, not in the future, but on an island, hypothetical, in the Indian ocean, not far from the Andamans, and inhabited by people who are descended from Buddhist colonists from the mainland.'[2] A similar intention runs through the remarkable stories of the American writer Ursula Le Guin. She places her ideal society in the distant future, when the descendants of the first space colonists have established the Ekumen. This galactic society of 3,000 nations spread across 83 planets is the macrocosmic community, the model of what our world could become by the power of education and the practice of self-enlightenment. The opportunity, Ursula Le Guin implies, is ours for the asking; and in *The Left Hand of Darkness* she describes a future time when men have to choose between the sovereignty of a planet and the higher order of a commonwealth of worlds. A prime minister tells the envoy from the Ekumen that fear of others and not love of country is the patriotism of Karhide:

> And its expressions are political, not poetical: hate, rivalry, aggression. It grows in us, that fear. It grows in us year by year. We've followed our road too far. And you who come from a world that outgrew nations centuries ago, who hardly know what I'm talking about, who show us the new road ... [3]

The reference to nationalism is typical of Ursula Le Guin's unusual gift for weaving the complex patterns of imagined societies into a mythic revelation for the twentieth century. No matter how far her imagination travels in future time or in deep space, she shares with the best science fiction writers a permanent concern for what the inhabitants of Terra have been doing, more often failing to do, since the earliest days of Cro-Magnon Man. In her most original story, *The Dispossessed* of 1974, she places the traditional utopian themes – self and community, poverty and wealth, socialism and capitalism – in a planetary location. There, in a way previously unknown to utopian literature but typical of late twentieth-century attitudes, Ursula Le Guin confirms the promise of the subtitle, *An Ambiguous Utopia*, by directing attention to the dilemmas in the classic systems of the ideal state. The story is presented as a search by the physicist of genius, Shevek, which begins in the hard conditions that govern life in the lunar colony of Anaress. The Anaresti have nothing and they have everything. Because they exist in the harshest imaginable conditions, they lack the most modest comforts of the technological utopia. Because they share in a gospel of equality and know nothing of profits, they live in a paradise of social concord and personal contentment. Perfection is no more than the highest point on a graph; and an essential ambiguity in the happy anarchism of Anaress reveals itself when Shevek has to leave his community in order to continue his search for a more complete knowledge in the prosperous, capitalist world of Urras. The puritan in Shevek condemns all the obvious evils in the self-seeking and frantic profit-making on that planet; and then another ambiguity deflects the narrative when the Terran ambassador tells Shevek to ponder the history of Earth: 'A planet soiled by the human species ... We controlled neither appetite nor violence; we did not adapt. We destroyed ourselves.' There is hope on Urras, because there it is possible to pursue knowledge and to reach a level of self-fulfilment not possible on Anaress. To the Terran ambassador Urras is close to paradise, even though 'it is full of evils, full of inhuman injustice, greed, folly, waste. But it is also full of good, of beauty, vitality, achievement. It is what a world should be! It is *alive*, tremendously alive – alive despite all its evils, with hope.'[4] Although the worlds of Anaress and Urras are separated, like the peoples of our planet, by different histories and different social systems, they have much to offer each other.

That is the point of the book, and the suggested harmonization of dualities and contraries is the ground of Ursula Le Guin's belief that 'realism is perhaps the least adequate means of understanding the incredible realities of our existence'. Through the fantasies of science fiction, she says, a writer 'may be talking as seriously as any sociologist – and a good deal more directly – about human life as it is lived, and as it might be lived, and as it ought to be lived'.[5] These remarks bring this discussion back to the origins of contemporary science fiction in the incredible realities of the post-war period, back to the beginnings of modern futurological practice in the desire to discover the most likely shape of coming things.

The many changes in world society during the last sixty years and, in particular, the swift acceleration in the rate of technological development since 1945 have imposed the conclusion that, in Peter F. Drucker's phrase, we are living in an Age of Discontinuity. This realization is the source of the most original science fiction stories and it is the starting point for the new practice of futurology. The fact that we may hope but cannot expect to go on living in the same old way has promoted a futuristic literature of allegories, anticipations, solutions and predictions that have universal significance. Thus, fiction and prediction are complementary activities of imagination and reason working upon, and going beyond, the contemporary experience of change. One of the best modern writers of science fiction, Brian Aldiss, says that his thinking began from the realization that 'the Bomb dramatized starkly the overwhelming workings of science and technology, applied science, in our lives ... '

> So I perceived, and have been trying to perceive more fully ever since, that my fiction should be social, should have all the laughter and other elements we associate with prosaic life, yet at the same time be shot through with a sense that our existences have been overpowered (not always for the worse) by certain gigantic forces born of the Renaissance and achieving ferocious adolescence with the Industrial Revolution.[6]

After the Bomb there came bigger and better bombs – the hydrogen bomb and Khrushchev's bomb that would rattle all the windows in Europe. And there were the computers, jet aircraft, nuclear submarines, television, space satellites, synthetic fibres, and the new industries that seemed to appear overnight – plastics, composite materials, electronics, atomic energy, micro-processing. The most spectacular demonstrations of the new rate of innovation were the Saturn V rockets of the American space programme in the 1960s. And yet, at the time when President Eisenhower signed the National Aeronautics and Space Act on July 29th, 1958, the Americans had only managed to put two small satellites into Earth orbit. Eleven years later, after a most intensive development of the new space technologies, 1,000 million viewers throughout the world looked in on that morning of July 16th, 1969, when Apollo 11 lifted off for the Moon at 09.32 hours Florida time precisely. Three days, four hours and fifteen seconds later, after a journey of one quarter of a million miles, the Lunar Excursion Module *Eagle* landed on the Sea of Tranquillity. 'The moment of touchdown was one of the moments of greatest drama in the history of man,' said the astronomer Sir Bernard Lovell. In the characteristic style of the 1960s he went on to speak of the future: 'The success in this part of the enterprise opens up the most enormous opportunities for the future exploration of the Universe.'[7] For any visitors from the Universe the astronauts left a message on the Moon:

HERE MEN FROM THE PLANET EARTH
FIRST SET FOOT ON THE MOON.
JULY 1969 A.D.
WE CAME IN PEACE FOR ALL MANKIND.

The message was the purest science fiction – the consummation of a hope that had grown with the advances in technology ever since Jules Verne first popularized the idea of the lunar journey with his story *From the Earth to the Moon* in 1866. The message to the Universe was a promise and an implicit prediction. It was the most recent stage in the growing expectation, first formulated by Winwood Reade in *The Martyrdom of Man* in 1872, that 'mankind will migrate into space, and will cross the airless Saharas which separate planet from planet, and sun from sun'.

Between those anticipations of space travel and the first steps of Neil Armstrong on the Moon a mere century had elapsed. That is the measure of the advances in technology and in state organization that have made it possible for men to leave their natural environment on Terra. A parallel process of development, unfortunately for the worse, has transformed the limited fire-power of the armies in the Franco-German War of 1870 into the world-destroying capacity of the latest intercontinental missiles. So, the world now alternates between hope and horror, between the vastness of space and the enormity of the weapons that have made some nations the lords of a last judgment. This absolute disjunction between past and present, which represents the most complete achievement of material progress, has generated an absolute demand for assessments of the moral and social consequences. According to Teilhard de Chardin we are moving towards the unification of all human societies. According to Marshall McLuhan we are changing our world into a global village. Thought and action are converging: 'Electric circuitry is Orientalizing the West. The contained, the distinct, the separate – our Western legacy – are being replaced by the flowing, the unified, the fused.' And the eminent zoologist Sir Julian Huxley commends the work of Teilhard de Chardin to English readers by saying that: 'We, mankind, contain the possibilities of the earth's immense future, and can realize more and more of them on condition that we increase our knowledge and our love.'[8]

To construe the immense changes of the present age is the special task of a vast new literature of the future. This grows in volume and continues to diversify in function as the prophets and the forecasters construct their models of the world in the next decade or in the next millennium. One area of exceptionally rapid growth is the new practice of futurology. The word seems to have been coined in 1943 by Ossip K. Flechtheim who was one of the first to argue for improvements in the descriptive methods that had dominated forecasts in the 1930s. At the end of the Second World War he was once again in advance of his time in seeking to promote university courses about the future on the ground that 'the last great obstacle toward an understanding of society is being removed. Instead of consulting the stars, the "futurologist" of 1945 can get his clews from historians and sociologists, from philosophers and psychologists, from political scientists and economists.'[9] The predictor was proved right, for that has been the way in which contemporary previsions of the future have evolved. And the prophet has been proved right in his view, remarkable for 1945, that the visions of utopian and dystopian fiction, 'daring and fantastic though they doubtlessly are, may yield insights that are more revealing than the voluminous writings of learned system-builders'.

The proof is the rolling barrage of dystopian fiction that opened with Orwell and Huxley in 1949 and did not moderate the furious premonitory onslaught until the end of the 1960s. And there is additional proof in the growth of postwar science fiction which communicates with a universal audience through books, films and television serials about all things possible in the marvellous or the menacing world of the future. Science fiction is the demotic literature of modern times. It has distinct affinities with pop music in so far as the participants are able to maintain close ties with the practitioners through clubs, magazines and annual conventions. It works through a continuing and self-adjusting mythology that ranges from absurd technological fantasies for the unthinking to moral parables of the evil that *Homo astronauticus* may spread throughout the galaxies. These stories are rooted in the conditions of their time and for that reason they are often an instant imaginative reaction to the most recent demonstration of the potentialities of modern technology. Although much of this fiction is no more than successful vogue-writing (like a large proportion of traditional novels) it has at least the merit of providing an indicator of the movement of contemporary moods; and these often find their realization, like the imaginary war stories of the decades before 1914, in visions of final disasters or of glorious triumphs.

One of the most frequent topics is the disaster story that liberates the mind from the anxieties of a troubled period by the convenient cancelling out of industrial civilization. It is indicative of the mood of the postwar years that this tale of the empty world, the private response of Richard Jefferies and W. D. Hudson to Victorian urbanization, has become one of the permanent pleasures of popular literature. And it is a pointer to the cosmopolitan roots of this stereotype that nowadays, in marked contrast to the exclusively European developments in this field after the First World War, the final catastrophe has caught the interest of many American writers. In fact, the success of the modern disaster story began with the appearance in 1947 of *Greener Than You Think* by the American writer Ward Moore. The Wellsian device of the fertility formula gone wild is the means of burying the world under grass. There is, however, a significant difference between Wells and Ward Moore in that urban civilization can no longer be saved. Everything vanishes before the green eruption, and the courteous American author arranges that England will be the last to go. God bless America:

> We heard the Queen and her consort remained in Buckingham Palace to the last, but this may be only romantic rumour. At all events, England is gone now, after weathering a millennium of unsuccessful invasions ... What few men of forethought who have taken to ships, what odd survivors there may be in arctic wastes or on lofty Andean or Himalayan peaks, together with the complement of the *Sisyphus* and its accompanying escort are all that survive of humanity ... The world has gone, vanished; but perhaps it is for the best, after all. We shall start again in a few days with a clean slate, picking up where we left off – for we have books and tools and men of learning and intelligence – to start a new and better world the moment the Grass retreats.[10]

The talk about 'a clean slate' is the customary text for the disaster stories. They are myths of reassurance. They carry the rainbow promise that *Homo sapiens* will face the evolutionary challenge in an exemplary manner. They are deceptive dreams. They pretend that the survivors of the great catastrophe will at last find a harmony and a meaning in their lives – once the complex, incomprehensible, frightening world of the twentieth century has disappeared. Thus, the American writer George Stewart ends a famous story about the obliteration of the cities, *Earth Abides*, with the last American musing on the future. As he looks at the New Ones, he thinks of the time when 'each little tribe will live by itself and to itself and go its own way, and their differences will soon be more than they were even in the first days of Man'. Stewart welcomes the happy return to the primeval state, and in a similar way the English writer J. G. Ballard finds pleasure in the disappearance of technology which he celebrates particular well in the decaying surrealistic landscapes of *The Drowned World* and *The Drought*. This profound and primitive pleasure, shared between writers and readers, derives from the feelings of hope or happiness that follow on the imagined annihilation of the contemporary world in stories with the appropriate titles of *Death of a World*, *The Death of Grass*, *The Great Calamity*. Once the survivors reach the zero point between the known past and the unknown future, then happiness begins with the contemplation of the return to the uncomplicated, primeval state of nature. In the variant version (in John Wyndham's *The Day of the Triffids*, for example) the hope of renewal lies in the empty cities which provide the means for restoring the human domination of nature. After the shipwreck of civilization the few surviving groups play the part of Crusoe. They salvage what they can from the ruins against the day when they will 'start a new and better world the moment the Grass retreats'. They prepare for the day when they 'will cross the narrow straits on the great crusade to drive the triffids back and back'.

Disasters continue to arrive and depart in contemporary fiction with a regularity that betokens the constancy of a dominant myth. At their centre, between the extreme assertions of the adolescent hope for a new start and of the apocalyptic urge to destroy the holy city of industrialism, there is a quiet zone of contemplation. The ending of all collective values in the breakdown of civilization is the opportunity to begin a search for the private values of selfhood that can only be found in the citadel of the individual mind. Thus, the violent pastoralism of *Heroes and Villains* by Angela Carter is a means of changing the future into the elemental world of the Soldiers and Barbarians who fight for possessions and women in an apocalyptic future of ruined cities and dense new forests. These provide the context for a study of love and hatred in the tribal period after the great disaster. The same *reductio ad humanum* is the key to *The Memoirs of a Survivor* of 1975, where Doris Lessing wipes out all the appearances of twentieth-century life in order to leave the central character completely free to discover the realities of the private self. It seems that the industrial city, the admired product of nineteenth-century progress, is nowadays seen to be both the cause and the symbol of painful uncertainties about the quality and the purpose of life in the twentieth century. The image of the ruined city is the expression of subconscious hopes and real fears; and the elements in this metaphor of endings and beginnings are now so universally recognizable that they carry over without translation to the visual idiom of television. In 1976, for instance, the B.B.C. presented a serial on this theme. The title was, of course, *The Survivors*.

These survival stories continue to find an audience because they belong with the dystopias and the communal utopias to the debate about inherited ideas and customary practices that has been going on since the Bomb. They are the most extreme statement in the argument against the imputed indifference and inhumanity of the industrialized society. The empty worlds of the future are the symbol of an absolute separation between humanity and technology, of a total disjunction between the industrial past and the inchoate future. The skeleton of death, visible in the ruined structure of the great city, is the image of the death-in-life that is destroying the instinctive and the natural throughout the world. The coming of a new life after the catastrophe is the common faith in these science fiction stories, which have very little to do with science. That fact contrasts with the fantastic applications of science and pseudo-science in the space travel stories that compose the favourite myth in the popular literature of our time.

At the beginning of the Kubrick film *2001: A Space Odyssey*, as mankind makes the great leap from Pleistocene to Palaeolithic, the camera cuts to a brilliant shot of the Orion spacecraft shining in the sunrise over Earth. The sounds of the 'Blue Danube' come in right on cue, as the spacecraft displays the marvels of technological progress and the sequence presents the order and harmony of the universe. These eloquent images are typical of the workings of modern science fiction. This popular literature is not a colloquy of soul with soul. That is the province of the traditional novel. As the frequent silent-running sequences in the Kubrick film demonstrate, science fiction is the building of images about the different relationships that decide our way of life in the environment of the Solar System. It is a literature of existences, not of essences; it is the one popular means by which an industrialized world can reflect upon the links between man and nature, between men and society, between the realities of today and the possibilities of tomorrow. Thus, the year 2001 is the occasion for a visual journey from the remote past – the origins once again – to the enigmatic suggestion of a new species emerging in the Star-child frames of the final sequence. The principal themes of the film (man–machine, time–space, progress–posterity) are part of the contemporary debate that runs through an alphabet of speculations about atomic bombs, cloning and demography to the xenobionts of a possible extraterrestrial zoology.

These speculations are the source material for a multitude of contemporary tales of the future; and an illustration of this link between the possibilities of modern science and the projections of futuristic fiction appears in an essay on 'A Crisis in Evolution' by the Director of the Kennedy Laboratories for Molecular Medicine at Stanford in California. After considering the rapid advances in the biological sciences, especially in experimental clonal reproduction, Professor Lederberg concludes:

> Meanwhile, a deeper understanding of our present knowledge of human biology must be part of the insight of literary, political, social, economic, and moral teaching; it is far too important to be left only to the biologists. In this spirit I can think of no better dedication than to the memory of the prophetic vision and artistic clarity of Aldous Huxley.[11]

Prophetic vision is too exalted a term for many science fiction stories which aim at little more than an exciting narrative of events in the future. The population problem, for example, is a matter that began to attract general attention after the first United Nations demographic projections and the forecasts of the hungry world of the future in the 1960s. These provide a natural topic for science fiction, and there are convincing accounts of overpopulation in Harry Harrison's *Make Room!*

Make Room! and John Brunner's *Stand on Zanzibar*. In fact Harrison has said that the writing of his book took him five years, 'just digging out the material to make an intelligent estimate of what life would be like in the year 2000 A.D. At this time there were no popular nonfiction books on the dangers of overpopulation, overconsumption, pollution and allied problems.'[12] This careful use of information is the strength and the weakness of science fiction stories. The painstaking experiments in creating imaginary societies, novel geographies and alien zoologies generally succeed in establishing the necessary fictional difference between then and now. And there too often the enterprise rests, because authors are content to follow the initial logic of their projections to incomplete conclusions. Their imaginary worlds remain the simple sum of the elements that suggested their composition; and their universal disasters fail to generate any adequate emotional response or significant psychological reaction in the bland narratives. For example, the accounts of jet-propelled infantry and a barbarous militarism are designed to give a futuristic context to the operations in Heinlein's *Starship Troopers*. In fact, there is no essential change. The ideas and the lethal weapons are the sentiments and the battle equipment of the Korean War enlarged to the cosmic dimensions of fantastic planetary worlds. In a similar way George Stewart manages to avoid dealing with the implied horrors in *Earth Abides* by giving his narrator the character of a dispassionate commentator, 'a student, an incipient scholar, and such a one was necessarily oriented to observe rather than to participate'.

The commercial requirements of the popular fiction market undoubtedly explain many a refusal to realize the true potential of a subject. The conventional models of science fiction are, however, a more frequent cause of the failure; for the authors learn from the tradition how they can control the greatest imaginable events – cosmic disasters or planetary discoveries – within the system of their fiction. The literary convention is that all things are proper subjects for rational explanation; and the literary practice is to give the appearance of intellectual order to the most haphazard and the most extraordinary events through the sensible report of a narrator or the impartial authority of an historical account. These differences in effect between the scale of the imagined experience and the limitations in style are most apparent in the homely and inadequate language of so many science fiction stories. The vocabulary of the street corner and of high-street shops provides words for the end of mankind or the arrival of the aliens from Alpha Centauri. 'One is annoyed', writes Stanislaw Lem, 'by the pretentiousness of a genre which fends off accusations of primitivism by pleading its entertainment character and then, once such accusations have been silenced, renews its overweening claims.'[13] The distinguished Polish writer of admirable planetary stories then goes on to criticize science fiction writers, especially the Americans, for their failings:

> By being one thing and purporting to be another, SF promotes a mystification which, moreover, goes on with the tacit consent of readers and public. The development of interest in SF at American universities has, contrary to what might have been expected, altered nothing in this state of affairs. In all candor it must be said, though one risks perpetrating a crime *laesae Almae Matris*, that the critical methods of theoreticians of literature are inadequate in the face of the deceptive tactics of SF.[14]

In considering Lem's critique of science fiction the theoretician of literature begins by noting that Lem achieved the apparently impossible in writing the two long and interesting volumes of his *Fantastyka i Futurologia* without a single mention of Zamyatin – the begetter of the modern dystopia and the Russian who dared to tell Stalin that for 'a writer to be deprived of the opportunity to write is a sentence of death'. Lem, however, leads a vigorous polemical life and is borne along by a crusader's faith in his fight for the New Jerusalem of a revelatory science fiction. Lem is a portent. He is the first writer of fiction ever to reflect at length (at considerable length) on the nature and purpose of fantastic literature. Moreover, he is one of the best of modern science fiction authors – the equal of Brian Aldiss, Philip K. Dick, and Ursula Le Guin – and he is likely to prove as good as H. G. Wells. His strength is that, like Wells, he had a scientific education and his thinking follows a strict dialectical line which is comparable to the clear-cut Wellsian ideology of social evolutionism. When he attacks the specious arguments of 'the apologists for the culture-shaping, anticipative, predictive and mythopoeic role of SF', his castigations are not envious but loving. He has the highest ideals for a form of fiction that is rarely serious enough for his ambitious conception of the illuminating and interpretational duties of a science-based literature.

He derides the mutants, robots, androids and other items in the stock-in-trade of popular writing. An uncompromising rationality tells him that science fiction should be the hypothesizing and creative workshop of a truly realistic literature. So, his major preoccupation is the anthropocentric prison in which terrestrial language and a brief planetary experience have confined the human race. At times he reads like a Marxist gloss on the Scholastic dictum that the intellect can conceive of nothing save that which comes to it through the senses. He labours to escape the imaginative pull of earth-type analogies; he sets himself the intellectual experiment of composing breakaway models of other kinds of existence in very different planetary worlds out there in deep space. The results can be examined in *The Invincible* (1973) in which human beings are given the problem of trying to understand a world where cybernetic organisms have evolved in ways that at first seem to be beyond terrestrial comprehension. In his earlier and more famous story, *Solaris* of 1961, Lem carries off one of the outstanding successes in the history of science fiction. He devotes a profusion of images and linguistic devices to the description of a planet that reflects the anthropomorphizing habits of *Homo sapiens*. The word *ocean*, for example, is the only means the explorers have for describing the plastic material that covers the entire surface of the planet Solaris. This ocean mirrors the operations of the human mind by producing forms and structures that seem to imitate the organic and inorganic shapes known to the terrestrial sciences. These are the *tree mountains*, *mimoids*, *fungoids*, *symetriads*, *assymetriads* and the other forms that the scientists have to classify in order to understand what they are. Can the Earthmen understand them? One scientist concludes that men are enclosed for ever in their own mental system and that 'there neither was, nor could be, any question of "contact" between mankind and any non-human civilization'. Another scientist concludes that a mystical communion with the Other, like that between the saints in revealed religion, is an eternal possibility.

These intuitions of an alien existence and the obsession with the incommunicable come naturally from a Pole. Lem finds himself in the frontier area of a planet divided between two great powers that have little communication with each other. Moreover, he shares in an historical memory of partition and domination by other nations. Although he abates nothing in his rigorous Marxist interpretation of literature, he belongs in his own singular way to the general movement of reappraisal. Differences are the staple of his fiction. Their importance for his understanding of the writer's task appears in his observations on the two extreme forms of modern science fiction – the world disaster and the interplanetary story. After saying that it would be interesting to have a realistic picture of the various reactions to the destruction of civilization, Lem concludes that 'this is not a subject for a science fiction writer, because the purely fictitious circumstances that bring about the disaster have no practical significance. It is rather a matter for the psychology of groups and individuals.'[15] This limiting of the science fiction story to an exclusive concentration on the data of the physical sciences is the choice of a severe rationalism that will not have anything to do with a baseless fantasy, nor with the making of myths and allegories. Now these may represent all that Lem does not want in science fiction; but the record shows that the Last Man, for example, has had a series of spectacular performances ever since Cousin de Grainville found in Malthusian theory and medical science the authority and the material for *Le Dernier Homme* of 1805. There is an area of the imagination that finds interest – solace, satisfaction or instinctive pleasure – in elegiac compositions about the end of things or in the attractive pastoralism of a world gone back to nature. To call this science fiction is immaterial, but it is a material fact that the tale of the future has always offered this opportunity of making moral statements about human wickedness and even of wickedly obliterating a hated world in some imagined catastrophe. Far below the Everests of pure reason the Sherpas may indulge their liking for play.

Lem is even more severe with the nonsensical science of the interplanetary story, and in particular with the Americanisms of an imaginary future that is too often a mish-mash of technological contraptions and meaningless adventures. Much of what he says about the artistic trivialities of American space fiction is just and apposite; and then one has to ask: Are his strictures necessary? Are all space stories trivial? The ephemeral frailties of popular literature are crushed in the iron embrace of a dialectical law that looks for meanings behind the appearances of fiction. It is true, for example, that the flying cities in the stories of James Blish are absurd inventions: spindizzies take New York through interstellar space, and antideath drugs keep the inhabitants alive for centuries. These contrivances, like teleportation and intergalactic wars, cannot bear the most cursory rational analysis; and when Lem fulminates about the banalities of transferring the class structures of Terra to the planets, the ephemeral evaporates in the white blaze of an incineration system.

The evidence shows that, from the crude beginnings of space fiction in the lunar travel stories of the Montgolfier period, the Solar System has represented a supreme challenge to humanity. Jules Verne gave the first technological answer in his vision of a world united (except for the British) in the glorious enterprise of sending the *Columbiad* to the Moon; and it was no accident that NASA gave the name of *Columbia* to the command module of the Apollo 11 Mission to the Moon. That political act was the acknowledgment of a real connexion between the imaginative imperative of the dream and the practical achievement of a rational objective. The action came naturally from the Americans. They are unique in their experience of the frontier life. They know that technology has reduced the vastness of a great continent to the human scale. They occupy the central position between the two greatest oceans of Earth. They are placed in a middle state between a European inheritance and a transatlantic development. Like the modern Russian state the American nation begins with a revolution that promises to change the future; and the Americans share with the Marxists the conviction that, in the Jeffersonian phrase, the future is to be a *novus ordo seclorum.*

The resolutely rational mind may declare that the proliferation of space travel stories in American fiction is irrelevant to serious literature. Nevertheless, the scale of production indicates a considerable and a popular industry in the work of many writers – Poul Anderson, Isaac Asimov, James Blish, Ray Bradbury, Hal Clement, Philip K. Dick, Ursula Le Guin, Robert A. Heinlein, Frank Herbert, C. M. Kornbluth, Keith Laumer, and so on to A. E. Van Vogt, Kurt Vonnegut, Donald Wollheim and Robert Zelazny. This extraordinary energy and the excellence of some writers (Ursula Le Guin and Kurt Vonnegut, for example) suggest that, just as the Europeans once found an ideal medium in the imaginary voyage, the American imagination has discovered an appropriate form in the space travel story. The mode reflects the experience of the westward advance of a new people who have made 'one giant leap for mankind' on the Moon. Space fiction permits the recapitulation of the achievements of the race; and for the serious it is a means of composing parables about modern man at a time when the world community is advancing towards the sixth millennium of civilized existence.

The contemporary idiom of *fans*, *fandom*, *fanzines* confirms the evidence of the last 200 years – that the tale of the future has always been the popular means of communicating the hopes and fears that follow out of the changing conditions of the great urban societies. In 1771 Sebastien Mercier presented his *L'An 2440* to the small literate world of the eighteenth century; and in 1870 Jules Verne produced his two greatest anticipations, *Round the Moon* and *Twenty Thousand Leagues under the Sea*, for an almost completely literate Atlantic community. In like manner Sir George Chesney's invention of the imaginary war story in the *Battle of Dorking* in 1871 was the appeal of a professional soldier to a nation. And the constant inventing of utopias and dystopias during the last 100 years is the continuing voice of a social conscience that speaks to the world with the optimism of Bellamy and Wells, the vehemence of Huxley and Orwell, the compassion of Hermann Hesse and Franz Werfel. The tale of the future has never been an affair of lonely hearts, not even on those occasions when the Last Man writes the final scenario for mankind. It is pre-eminently a product of the social imagination. It is a communal activity, a dialogue in which writers lead the discussions about the future of the nation in a warring world or about the structure of society in a co-operative world. The tale of the future was invented and has been developed in order that it can be all things to all imaginations. As it was in the beginning, so it is now the dream-time of urban civilization and – to mix an American metaphor – it is a stamping ground for the latest thoughts on the citizen and society, on the magnificent promises and menacing possibilities of the applied sciences, and on all those other matters that may decide the pattern of life on Terra in the third millennium of the Christian era.

We live in the Heracleitan Age. We exist, as Hannah Arendt has it, between past and future. We know with the absolute conviction of profound experiences that tomorrow will be different from today. Urban guerrillas, the Iron Curtain, the Hot Line and the Ecumenical Movement – the many great changes of the last thirty years repeat the Heracleitan message that 'all is flux, nothing is stationary'. The prophets and the predictors of the late twentieth century press forward with the examination of social and scientific possibilities that began with Mercier and Malthus in the late eighteenth century. This is to say that the powers of technology and the growth of population have been the primary regulators of urban expectations from the days when Watt's steam-engine and Jenner's vaccine applied the discoveries of science to the benefit of mankind. Ever since the momentous innovations of the first industrial revolution the mind has sought with an ever-increasing sense of urgency to anticipate all the consequences of the perpetual flux by creating patterns of expectation. There is no end to the modelling of possible future worlds. The techniques of the forecasters become ever more refined and the images of the future compose a familiar code for the television sets of the world. Indeed, the exceptionally rapid development of futurology during the last two decades shows that, although the prophets may now be honoured in their own countries, the predictors have become the mentors of governments everywhere.

Futurology is an ugly, pretentious word. It is, however, the best available means of naming the present stage in the evolution of the informed, professional estimate of future developments that began with the sociologists, town planners and the military in the last quarter of the nineteenth century. Futurology is yet one more advance in the specialization of intellectual labour. Like the witches in *Macbeth* the modern forecasters have been summoned by the necessities of state – by the demand for a show of coming things that may provide the rationale for political action. Government now speaks like Macbeth:

> I conjure you, by that which you profess,
> Howe'r you come to know it, answer me ...

Most of the answers come, as Ossip K. Flechtheim predicted in 1945, from those who have professional knowledge of a science or a social science – the anthropologists, political scientists, sociologists, psychologists, economists. The applications of their knowledge began to influence governments in the 1930s, when the theories and advice of economists (Keynes, for example) were a guide to politicians who had to deal with the problems of unemployment. The expert became a normal source of information for the planning staffs during the Second World War, when the role of the scientific adviser (Lord Cherwell with Churchill, for example) had at times a crucial importance. Some of the most important decisions on the planning of future operations began with the work of the Operational Research Sections; and undoubtedly the most fateful decisions in the war against the U-boats turned on the calculations and predictions of a small group of some sixteen scientists in the Coastal Command of the Royal Air Force. One of them was C. H. Waddington, who later became the Professor of Animal Genetics in the University of Edinburgh. In his history of their work he writes that it was 'simply the general method of science employed to study any problem which may be of importance to an executive'. Their collaborative methods were the model of what is now a recognized system of planning future actions:

> There was, for the U-boat war alone, the whole apparatus of a professional science – the regular journals, the handbooks, the classical papers, the theoretician's approach, the experimentalist's approach, the whole complex structure which gave a meaning to the details which could be fitted in to the picture.[16]

The successes of the U-boat war were a lesson for governments engaged in planning the reorganization of social and industrial life after the war. So, Waddington became a propagandist for the use of Operational Research methods, and in the September of 1946 he stated his beliefs:

> There is very little doubt that this type of science will be widely used in the future in many different fields. Men with experience of the wartime developments are in fact already at work in various Ministries – such as Housing, the Board of Trade, etc. – in jobs which, whatever they may be called, are in fact operational research assignments. There are many more places in which a similar approach would be valuable. It is one of the main purposes of this book, by expounding what operational research did in one particular field, to suggest what kinds of things it might be called upon to do in others.[17]

Although those words reached the galley-proof stage in 1946, the book had to wait on the Official Secrets Act until permission to publish was at last given in 1973. In the meantime Waddington continued to argue, good scientist that he was, for a rational and humane method of preparing for the changes that never cease to come. One of his last acts before his death was to serve as chairman to the first symposium ever held in the United Kingdom on the matter of *The Future as an Academic Discipline* which the Ciba Foundation arranged in the February of 1975. His opening remarks are a summary of the debate about the future that began some thirty years before:

> We are here to discuss whether universities in general and British universities in particular should take account of the problems that mankind is obviously going to face in the next few decades, and, if universities are to do this, how should

> they do it? It is only because the situation in the next few decades is clearly going to be unlike what it was in our grandfathers' days that I think the question arises so seriously now. We all recognize that we are facing a series of crises which can't be completely separated from one another. Each one of them – atomic warfare or the population problem or the environment problem or the energy problem or what have you – is a considerable threat.[18]

The unconditional imperative of a willed adaptability repeats endlessly and profusely through a vast new literature the one message of intelligent anticipation. Futurology has developed at such a speed – from occasional books in the 1940s to national institutes in the 1970s – that it must represent the swiftest advance in the organization of knowledge since the explosion of publications after the appearance of *The Origin of Species* in 1859. Today the earliest statements of the predictors read more modestly than many of the sweeping claims of recent years. In 1941, for example, S. and R. Rosen were cautious in describing the state of forecasting in the introduction to their *Technology and Society*: 'Investigation of the effect of the machine upon society has only recently begun to claim the attention of writers.' In 1942 the Australian economist Colin Clark published one of the most important books in the early period of modern forecasting. In his introduction to *The Economics of 1960* he described specific techniques that still apply to forecasting in the field of economics:

> The genesis of this book lay in the Author's realisation that he, like all other practitioners of applied economics, was continually having to take steps which involved estimates of the economic situation at dates far into the future. Confronted by this predicament, one has a natural tendency to assume that trends which have been in existence during the recent past will persist into the future. In many cases it can be shown that policies based on such assumptions are almost sure to be wrong. In this book an attempt is made to deduce, by fuller analysis, the most probable course of world populations, industrial development, prices, capital movements and interest rates over the next twenty years.[19]

The developments in forecasting during the last thirty years are the story of a continual advance in the methodology of the subject and in the diffusion of the latest ideas about the future. During the 1940s and the 1950s the usual style in predictive writing was the descriptive account for the intelligent reader which had been the norm before the war. Thus, the various authors who contributed essays to *The Prospect before Us: Some Thoughts on the Future* (1948) wrote 'in the belief that man is at a cross-roads and that his present decisions will have profound and far-reaching consequences. If the general outline of the future could be delineated, it might enable men to take more enlightened action with regard to contemporary dilemmas.'

This talk of cross-roads and consequences, so typical of the postwar tale of the future, is the starting point for the discourse of the forecasters. From 1950 onwards a new generation of path-finders sets out to discover the trends in world society that may decide the condition of life in the last decades of the twentieth century or in the more distant future. The range of their investigations is evident in two books of 1952. In *Die Zukunft hat schon begonnen* the German writer Robert Jungk looked at the technological innovations then being introduced, and he directs attention to the ways in which society will have to adapt to the changes that will come. The title of *The Next Million Years* calls for evolutionary explanations on the Darwinian scale, and they come in the most appropriate way from Sir Charles Galton Darwin, a distinguished natural scientist and the grandson of Charles Darwin. He tells the layman that his method is to make general deductions about the future of mankind on the analogy of the behaviour of the molecules in conservative dynamical systems. He is thinking in particular of Boyle's Law which, relates pressure to volume:

> When I claim that we ought to be able to foresee the general character of the future history of mankind, I am thinking of this analogy. The operation of the laws of probability should tend to produce something like certainty. We may, so to speak, reasonably hope to find the Boyle's Law which controls the behaviour of those very complicated molecules, the

> members of the human race, and from this we should be able to predict something of man's future. It is not possible to get something out of nothing here any more than it is in the case of the gas; so the possibility depends on finding out whether there are for humanity any similar internal conditions, which would be analogous to the condition of being a conservative dynamical system, and external conditions analogous to the containing vessel. If both these demands can be satisfied, then there is the prospect that a great deal can be foretold of the future of the human race, and this without any very close detail in the basic principles from which it is derived.[20]

The prospects of the molecules are not good. As Sir Charles Darwin points out, the fuel supplies are beginning to run out, and the alternative sources of energy in atomic power and water power do not hold out hopes that men will continue to win energy from nature on the scale to which they are accustomed. Further, the food supplies cannot keep pace with the growth of population; and even if mankind contrives to increase the production of food, the iron laws of biology tell the author that 'there will always be exactly the right number of people to eat it. It all comes back to Malthus's doctrine and to the fact that an arithmetical progression cannot fight against a geometrical progression.' Finally, the limitation of births is a practice that works against the best interests of the more prosperous nations. Since a stationary population does not enjoy the tonic effects of natural competition, it must gradually degenerate. It is, therefore, 'impossible to believe that a degenerating small population can survive in the long run in a strongly competitive world, or that it can have the force to compel the rest of the world to degenerate with it.'[21]

These observations mark the extent of the great changes in expectation that have taken place since the more self-confident decades of the last century. In the 1840s and the 1850s, when the Prince Consort and the Poet Laureate spoke about the better things that would come for the greatest industrial nation of that time, they used the new language of progress – historical developments, social improvements, the march of mind, the ringing grooves of change, the delights of 'living at a period of most wonderful transition'. The conviction of change is stronger than ever today, but the wonders of transition no longer appeal to as many as they once did. The main effort in the forecasts of the last thirty years is eminently prudential. The most frequent intention is to describe the changes-to-come so that society can decide on the best ways of preparing for, or of guarding against, the developments that may decide the future. Although the method of these projections still works from the old linear assumption of a constant rate of advance, there is a new eagerness to maintain the continuities between present and future. By the 1970s this concern for the future had evolved into a new species of predictive writing in a series of books and articles about the various changes that are expected to affect the well-being and even the continued existence of the great technological societies. This latest response to the age of crisis, as the new literature likes to call it, communicates a multitude of anxieties in a hideous jargon of *future shock*, *rethink*, *para-primitive*, *overchoice*, *survival strategies*.

The virtue of self-preservation is the lowest common morality in these warnings. Scientists, philosophers, economists, sociologists, and journalists join in telling society that the future will be exactly what the world deserves. The index to any of their books displays the new catechism of collective responsibilities: anticipation, assessment, atomic energy, birth rates, computers, death rates, energy crisis, famine, Gross National Product, hydrogen bomb and so on to wind power, young dropouts and zooplankton as food. At their best the new forecasters are the old moralists brought up to date. Thus, the American geochemist Harrison Brown ends a most perceptive and influential analysis in *The Challenge of Man's Future* (1954) by giving the world three choices: either a reversion to an agrarian existence, if war is not abolished and populations are left uncontrolled; or 'the completely controlled, collectivized industrial society'; or 'the world-wide free industrial society in which human beings can live in reasonable harmony with their environment'. His last words give a generous emphasis to the moral obligations of his own people, as he sees them:

> We in the United States are in a position of overwhelming responsibility at the present time, for in a very real sense the destiny of humanity depends upon our decisions and upon our actions. We still possess freedom, our resources, and our knowledge, to stimulate the evolution of a world community within which people are well fed and within which they can lead free, abundant, creative lives.[22]

In another book of 1954, *La Technique ou l'enjeu du siècle*, a comparable plea for the reconsideration of aims and objectives in the technological society comes from Jacques Ellul, professor in the Faculty of Law at Bordeaux and a hero of the French Resistance. He begins by saying, in the preface to the revised edition, that 'in the modern world the most dangerous form of determinism is the technological phenomenon. It is not a question of getting rid of it, but, by an act of freedom, of transcending it.' After examining the effects technology has upon society, he ends by asserting the moral priority of ends over means. He asks that men should make 'the effort to discover (or rediscover) a new end for human society in the technical age. The aims of technology, which were clear enough a century and a half ago, have gradually disappeared from view.'

The same appeal sounds through the utopias and the dystopias of the postwar period. Despite their differences of mode and method the prophets and the predictors are at work in the one field of contemporary experience. Thus, the variety of the expectations in the tale of the future—from dystopias to self-confident space fiction—is reflected in the range of predictive writing. The forecasts advance from the dangers or disasters that await mankind to hopeful accounts of an increasingly progressive and peaceful world society. At one extreme there is the evolutionary theorist, Sir Charles Galton Darwin, who argues in *The Next Million Years* that mankind cannot cheat nature; and his many successors continue to this day to publish their worst fears for humanity. Their titles show that their main interests are in the dangers of modern warfare, environmental pollution, world population, new biological developments and urban problems: *Unless Peace Comes*, *Kill and Overkill*, *Planet in Peril*, *Before Nature Dies*, *Expanding Population in a Shrinking*

World, The Biological Time Bomb, The Private Future, An Inquiry into the Human Prospect. By contrast there are many other writers, scientists and journalists in particular, who display a happy faith in techniques and are confident that human reason like a cunning steersman will pilot the way through to a more stable future. One of the earliest and most striking illustrations of this robust attitude of mind appears in *The Foreseeable Future* of 1957. The author was the eminent scientist Sir George Thomson, Nobel Laureate and Master of Corpus Christi College, Cambridge. In the customary manner of these publications he examines all the developments that promise to affect the future of mankind, and he ends on a note of reassurance:

> Civilized mankind at present is rather like a child who has been given too many toys for its birthday. Life seems unreal and out of focus. Shall we be able to settle down or will there be a constant and bewildering succession of fresh toys? Even allowing for the disadvantage of a very large population and of the exhaustion of oil and of the easily accessible supplies of metals, it is reasonable to expect that expanding knowledge and advanced technology will allow a steadily improved standard of living.[23]

Expansion, advance, improvement—these have always been the major certainties in the idea of progress. They are now the slogans of the optimists who seem able to separate the accumulation of knowledge from its applications and the growth of technology from the many consequences that follow on the use of new methods and new materials. Winwood Reade could have written his last paragraph for the Nobel Laureate:

> There is no reason to anticipate that anything irreparable will go wrong with the earth physically for many millions of years, and are there not other planets and other stars? It is difficult to exterminate a species once well established, and man's best efforts to kill himself are unlikely to be more successful than those of the plague bacillus or the influenza virus. Even with the present brains of intelligent people Man may expect a glorious future. Who will dare to set limits to what he may reach as his brain improves? This future is not foreseeable![24]

The popularizing of technological possibilities joined together many writers in many countries in the happy marriage of identical minds. American, British, French, German, and Russian publications show a rare unity of belief in forecasting great advances for the industrial nations. Their common faith is a comfortable materialism that equates scientific progress with social improvement; and their general method is to look for their information in interviews with scientists or in the latest research papers. The Russian authors of *Life in the Twenty-first Century* (1959) visited 29 Soviet scientists in the course of assembling their material; and the American author George Soule gathered most of his ideas for *The Shape of Tomorrow* (1958) from a survey of the effects of technological changes made by the Twentieth Century Fund. Wherever the Russians and the American touch on the same facts, they agree so completely about the shapes of things-to-come that their forecasts, but not their ideologies, are interchangeable. This popularizing activity knows no boundaries, and on occasions the research paper of today becomes an item in the book of tomorrow. On August 23rd, 1974, a short report appeared in *Nature* on the feasibility of constructing artificial habitats in space, where 'communities could be as comfortable as the most desirable parts of the Earth, with natural sunshine, controlled weather, normal air, apparent gravity and complete freedom from pollution'.[25] The author is the Professor of Physics at Princeton University, and in the September of 1974 he published a longer article on space colonies in *Physics Today*. That provided the information for a commentary in *New Scientist* for October 24th, 1974; and all three articles were quoted in support of the chapter on 'Flying City-States' in the 1976 edition of *The Next Ten Thousand Years* by the British writer Adrian Berry.

This natural progression has been a constant in the rapid growth of forecasting since the end of the Second World War. After the initial phase of popularization in the 1950s, the 1960s were a period of exceptionally rapid development and of increasing specialization. It was the time of the first journals, the first international conferences, and the first institutes of the future. These were the major indications of a movement that went on, and still continues, throughout the industrialized nations; and in consequence it is nowadays unusual to find a country in which this prospective attitude of mind does not influence the planning committees and the national development boards. The interests of nations, of commerce and industry coincided in the search for more reliable predictive techniques in the fields of economics, the social sciences and the applied sciences. The new journals filled with articles that addressed their readers in the language of operational codes, logistic curves, systems analysis, trend extrapolation, morphological research, relevance tree techniques. At the same time there was a succession of most influential and original publications in which writers applied their expert professional knowledge to the discussion of probable developments. For example, in 1963 three experts collaborated in producing *Resources in America's Future*, which was devoted 'to the projection of demand and supply of natural resources – their products and services as well as the basic land, water, and minerals – to the year 2000 for the United States.' In 1964 Robert Jungk began a series of 15 books, *Modelle für eine neue Welt*, which were designed to examine the entire range of possibilities from the future of world population to urban society and the 30-hour week. And in 1964 the British journal *New Scientist* published a series of some hundred articles in which experts gave their views about the future. The editorial introduction was a statement of cause and effect for an audience of scientists:

> As far as I know, no journal has ever attempted to forecast the future in this way or on this scale, and a word of explanation should be given. There is a growing awareness that rates of change are now so great that medium-range forecasts are a serious requirement if we are not to be caught out by change and if the scientific revolution is to be carried through wisely. But isolated forecasts by individuals tend to centre on a few ideas and overlook the fact that the world of the future, like the world of the present, is a complex system of many technical and human elements. A comprehensive approach is required if important factors are not to be missed.[26]

In 1965 the Organization for Economic Co-operation and Development commissioned Erich Jantsch to study the methods and the uses of technological forecasting in their member countries. 'The most important message', he reported, 'to emerge from the current surge of interest in technological forecasting is perhaps this: our ability to choose which technological developments we wish to accelerate means that we can mould our own future.' That belief was repeated in the introduction that Professor Daniel Bell, Chairman of the Commission on the Year 2000, wrote for *The Year 2000* (1967) by Hermann Kahn and Anthony J. Wiener. In his view the most important reason for the growing interest in the future was

> ... the simple fact that every society today is consciously committed to economic growth, to raising the standard of living of its people, and therefore to the planning, direction, and control of social change. What makes the present studies, therefore, so completely different from those of the past is that they are oriented to specific social-policy purposes; and along with this new dimension, they are fashioned, self-consciously, by a new methodology that gives the promise of providing a more reliable foundation for realistic alternatives and choices, if not for exact prediction.[27]

The British version of that theory appeared in 1968 in *Forecasting and the Social Services*, where Michael Young argued that 'the future will be largely shaped by the choices men make, or fail to make, and not simply by technical forces; that processes existing and apparent now can reveal some of the basic choices that will confront men over the next thirty years; and finally, that social science should not only provide tools (trained personnel, institutions, theories and methods) but also help men to extend their visions.'[28]

The social sciences are, however, only a part of the new intellectual industry of pattern-recognition and trend-detection. The many innovations of the last twenty years show that the exploration of the future is a favourite obsession of modern times. The future is the one question mark that engages the interest of all our world; and at this point in the last pages of a last chapter it is appropriate to raise questions about the value and the validity of an intellectual activity that has grown in complexity and in importance since the days of Malthus and Condorcet. The first answer is that, no matter what may be said about the methods or the conclusions of many forecasts, there can be no doubt about the moral energy and the great seriousness of the futurologists. The best of them display an admirable commitment to the many problems of a technological epoch. There is the exemplary work of Dennis Gabor, the wise and humane scientist who invented holography. In 1963 he published his original reflections on the subject of *Inventing the Future*, and his message for the anxious was to affirm the sovereignty of the human will: 'We are still the masters of our fate. Rational thinking, even assisted by any conceivable electronic computers, cannot predict the future.' In his next book, *Innovations: Scientific, Technological, and Social* (1970), he argued for order and proportion in all innovations; and he suggested a number of social devices which could serve to modify random behaviour in a technological world.

The counterpart of the scientist philosopher is the scientist theologian – Pierre Teilhard de Chardin. Like a Wellsian time-traveller he advances through evolutionary theory and religious doctrine to the stupendous conclusion that 'the *End of the Species* is in the marrow of our bones!' He develops this theme in three major studies, *The Phenomenon of Man*, *Man's Place in Nature* and *The Future of Man.* There the Jesuit palaeontologist reflects upon the histories of man and nature. He sees them as a convergent movement, as advances towards the Omega Point, the last point of arrival in the cosmic convolution that began with man's conquest and organization of Earth. As the human atoms press ever more closely together within the molecular structure of mankind, the constant growth of knowledge ensures the unification of our world. The development of the new electronic machines will do for the brain what the invention of the telescope did for human sight. As the capacity for thought increases, mankind will move forward in step with a self-reflective evolution that will discover the salvation of the thinking species outside the limitations of Time and Space. The end of the world will be the fulfilment of human biology and the beginning of a new spiritual evolution:

> Pressed tightly against one another by the increase in their numbers and relationships, forced together by the growth of a common power and the sense of a common travail, the men of the future will in some sort form a single consciousness; and because, their initiation being completed, they will have measured the power of their associated minds, the immensity of the universe and the narrowness of their prison, this consciousness will be truly adult, truly major. May we not suppose that when this time comes Mankind will for the first time be confronted with the necessity for a truly and wholly human act, a final exercise of choice – the yes or no in face of God, individually affirmed by beings in each of whom will be fully developed the sense of human liberty and responsibility?[29]

After the sweep of that prediction – the most ambitious in the history of futuristic literature – it is only possible to answer the earlier question about the validity of futurology by saying that time will tell. In fact, time has already shown that the urge to foresee the future is at once a primal curse of human existence and the condition of a safe passage through the turbulent constraints of a technological epoch. The evidence of the last 200 years, from Turgot to Teilhard de Chardin, reveals the essential dualities of the rational animal. We are caught between the static and the dynamic, between the lived experience of the past and the uncertainties of the present. We desire the permanent repose of some final attainment, the illusion of finality, but we are committed to action by the nature of our infinite desires. Tennyson puts the dilemma in this way:

> Let us alone. What pleasure can we have
> To war with evil? Is there any peace
> In ever climbing up the climbing wave?
> All things have rest, and ripen towards the grave
> In silence; ripen, fall and cease:
> Give us long rest or death, dark death, or dreamful ease.[30]

The prophecies and the predictions of the last 200 years show that the future is not accurately predictable on any scale nor in any detail. There are too many variables and no one can guess what changes may divert the apparent direction of human development. It is a fact, however, that we cannot do without the prophets and the predictors. We need them as we need the air we breathe; for they offer the only means by which individuals and nations can find their directions in a changing world. The proliferation of utopias and dystopias in the last 30 years is the work of the moral imagination intent upon asserting the right conduct of human affairs in a time when the abuse of power – political and technological – threatens the continued existence of society. The growth of social and technological forecasting since the 1950s is the rational, collective solution to the many organizational problems that come with constant change. The importance of these innovations is in the way the forecasters focus attention on the problems of ends and means. Their greatest service is in spreading the word that we must choose today in order to have the tomorrow we desire.

Finally, it is just and fitting that this writer should now face the judgment of the future. He forecasts that, as the evidence of futuristic literature demonstrates, we are moving out from the European epoch of human history at the same time as the other great human groupings are beginning to advance from the continental periods of their separate social evolution. Our immediate future is to be shared between the Europeans and the peoples of the Americas in the Atlantic epoch of an international development. The steady rise in the level of education everywhere and the growth of communications will accelerate the movement towards a more united planet. When the generation of the Second World War has gone from power, we can expect that their successors will perceive the advantages of developing more effective international institutions. But there the prediction has to end in the simple hope that the known process of political evolution – from tribe to monarchy to federation – will continue to operate in the twenty-first century.

And this book has to end where futuristic literature had its popular origins – with the French. Although the Red Man has not yet recovered his ancestral lands in North America, as Sebastien Mercier foretold that he would, a referendum of the British people has realized the dream of a European community described in *L'An 2440*. Looking backward for the last time at that futuristic utopia of 1771, it must be apparent that prophecy and prediction offer an enigmatic encouragement to the human race. Mercier described some of the changes that came with the French Revolution; but he never foresaw how his desired improvements would come about. He learnt by the hard experience of 1792 that his country had to choose a future for itself. In that fateful year, when the future of the young Napoleon was still unpredictable, the French found themselves between the ending of their ancient monarchy and the hazardous beginning of the new republic. On July 19th, when the Austrian and Prussian armies under the command of the Duke of Brunswick crossed the frontier, everything pointed to the certain defeat of the revolutionaries. On September 2nd, when Verdun surrendered to the invaders, it seemed that nothing could halt the advance on Paris and the restoration of the monarchy. A number of recent accidents had, however, placed Danton in the Ministry of Justice, and on September 2nd that great orator called on the people to decide their future in a famous speech to the Legislative Committee of General Defence. His one remembered slogan is an apt epigraph for futuristic literature: *De l'audace, et encore de l'audace, et toujours de l'audace!*

Notes

NOTE: Wherever a later edition of a book has been used for reference the date of the first edition is also given in parentheses.

Prologue: The Idea of the Future

1 Quoted by A. C. Crombie in *Augustine to Galileo*, 1961, v. 1, p. 55.
2 William Rawley, 'To the Reader', *The Works of Francis Bacon*, 1803, v. 2, p. 1.
3 Joseph Glanvill, *The Vanity of Dogmatizing*, 1661, pp. 181–2.

Chapter 1: The Discovery of the Future

1 Samuel Madden, *Memoirs of the Twentieth Century, being Original Letters of State under George the Sixth*, 1733, p. 136.
2 Anonymous, *The Reign of George VI, 1900–1925*, edited by I. F. Clarke, 1972, p. 23.
3 For a full list of futuristic stories see: I. F. Clarke, *The Tale of the Future*, Library Association, 1978.
4 Ronald L. Meek, *Turgot on Progress, Sociology and Economics*, 1973, p. 59.
5 Sebastien Mercier, *Astraea's Return; or the Halcyon Days of France in the Year 2440*. Translated from the French by Harriot Augusta Freeman, 1797, pp. 113–14.
6 Sebastien Mercier, *The Waiting City. Paris, 1782–88*. Translated and edited by Helen Simpson, 1933, p. 314.
7 Julian P. Boyd (ed.), *The Papers of Thomas Jefferson*, 1953, v. 7, p. 136.
8 Peter Cunningham, *The Letters of Horace Walpole*, 1906, v. 8, p. 438.

9 Wilhelm Kurrelmeyer, *Wielands Werke, Prosaische Schriften II*, 1783–94, 'Die Aeropetomanie', p. 1.

10 A. Condorcet O'Connor and M. F. Arago, *Oeuvres de Condorcet*, 1847, v. 6, p. 626.

11 A. Gordon, *An Historical and Practical Treatise upon Elemental Locomotion by Steam Carriages on Common Roads*, 1832, p. 37.

Chapter 2: To the Last Syllable of Recorded Time

1 Sebastien Mercier, *L'An deux mille quatre cent quarante*, 1787, v. 2, p. 190.

2 Joel Barlow, *The Columbiad*, 1809, pp. 366–7.

3 *Ibid.*, p. 357.

4 Robert Southey, 'A Tale of Paraguay', Canto 3, *Poems of Robert Southey*, edited by Maurice H. Fitzgerald, 1909, p. 661. For a comparable attitude to the powers of science see: William Jackson, *The Four Ages*, 1798, where the author writes (pp. 82–3) that 'the progress towards perfection may be seen in the face of the country and the appearance of the towns ... As the poets formed a Golden Age, according to their imagination of what is good or desirable, I may, in my turn, imagine what will be the situation of mankind, when genius, corrected by science, and assisted by reason and virtue, shall have produced that improvement of society to which it naturally aspires – this is the millennium of philosophy.'

5 T. R. Malthus, *An Essay on the Principle of Population* (1798), Everyman's Library, 1960, v. 1, p. 8.

6 William Hazlitt, 'Mr Malthus', *The Spirit of the Age* (1825), World's Classics, 1960, p. 160.

7 Mircea Eliade, *Myth and Reality*, 1964, p. 59.

8 Anonymous (Jean-Baptiste François-Xavier Cousin de Grainville), *The Last Man; or Omegarus and Syderia, A Romance in Futurity*, 1806, v. 1, p. 21.

9 *Ibid.*, p. 141.

10 *Ibid.*, p. 88.

11 Julius von Voss, *Ini*, 1810, p. 1.

12 Derek Roper (ed.), *Lyrical Ballads 1805*, 1968, pp. 35–6.

13 Anonymous, *A Hundred Years Hence; or the Memoirs of Charles, Lord Moresby, written by Himself*, 1828, p. 109.

14 Baron Georges de Cuvier, *Recueil des éloges historiques*, 1819, v. 1, p. 20.

15 'Report of Dr. Ure's Lecture on the Steam Engine in aid of

the Funds for erecting a Monument to James Watt', *The Glasgow Mechanics' Magazine*, 1825, v. 2, p. 384.

16 Dionysius Lardner, *The Steam Engine, Steam Navigation, Roads and Railways* (1827), 1851, p. vi.

17 Quoted by W. H. G. Armytage in *A Social History of Engineering*, 1961, p. 124.

18 Charles Gide, *Design for Utopia. Selected Writings of Charles Fourier*, 1971, p. 180.

19 Jane Webb, *The Mummy*, 1827, pp. 201–2.

20 Johann Peter Eckermann, *Conversations with Goethe*, Everyman's Library, 1930, pp. 173–4.

21 Lord Macaulay, 'Southey's Colloquies on Society', in *Literary Essays Contributed to the Edinburgh Review*, 1932, p. 133.

22 James Patrick Muirhead, *Historical Éloge of James Watt by M. Arago*, 1839, pp. 149–50.

23 J. A. Etzler, *The Paradise within the Reach of All Men*, 1836, 2nd Part, p. 213. See also: *The New World; or a Mechanical System to perform the Labours of Man and Beast*, 1841; and *Emigration to the Tropical World*, 1844. He begins the latter by saying: 'We are on the eve of the most eventful period of mankind. Migrations of millions from north to south will soon take place, and new nations and empires will be founded by them, superior in every respect to any known in history.'

Chapter 3: The Prospect of Probabilities

1 William N. Griggs, *The Celebrated 'Moon Story'*, 1852, p. 49.

2 *Ibid.*, pp. 78–9.

3 *Ibid.*, p. 95.

4 *Ibid.*, pp. 98–9.

5 John H. Ingram (ed.), 'Richard Adams Locke', in *The Collected Works of Edgar Allan Poe*, 1890, v. 4, p. 489.

6 Sir John Herschel, *A Preliminary Discourse on the Study of Natural Philosophy*, 1831, pp. 49–50.

7 Quoted by Pierre Versins in *Encyclopédie de l'utopie, des voyages extraordinaires, et de la science fiction*, Lausanne, 1972, p. 93.

8 Félix Bodin, *Le Roman de l'avenir*, 1834, pp. 2–3.

9 *Ibid.*, p. 20.

10 *Ibid.*, pp. 54–5.

11 *Un Autre Monde par Grandville* (1844), 1963, pp. 116–17.

12 John H. Ingram (ed.), 'The Balloon Hoax', *op. cit.*, v. 1, p. 94.

13 *Illustrated London News*, January 6th, 1849, p. 9.

14 D. Ryzanoff (ed.), *The Communist Manifesto*, 1930, pp. 31–2.
15 Michael Angelo Garvey, *The Silent Revolution*, 1852, p. 1.
16 *Ibid.*, pp. 170–1.
17 *Ibid.*, p. 179.
18 Lord Macaulay, 'Lord Bacon', *Literary Essays*, 1932, p. 385.
19 Henry Thomas Buckle, *History of Civilization in England* (1857–61), 1885, v. 1, pp. 140–1.
20 *Ibid.*, p. 807.
21 *On the Progress of Mankind and Reform* by the Honourable George Bancroft, late U.S. Minister in London, 1858, p. 30.
22 Charles Darwin, *The Origin of Species* (1859), 1912, p. 412.
23 *Ibid.*, p. 412.
24 Anonymous (E. Burrows), *The Triumphs of Steam*, 1859, pp. 1–2.
25 Andrew Park, *The World, Past, Present and Future*, 1862, p. 17.
26 William Ellis, *Thoughts on the Future of the Human Race*, 1866, pp. 54–5.
27 Edmond About, *Le Progrès*, 1864, p. 31.
28 Karl Vogt, *Lectures on Man*, 1865, p. 156.
29 Ludwig Büchner, *Man in the Past, Present and Future*, 1872, p. 163.
30 Thomas Henry Huxley, 'Man's Relations to Lower Animals', in *Man's Place in Nature and other essays*, Everyman's Library, 1906, p. 103.
31 Charles Knight, *Passages of a Working Life*, 1865, v. 3, p. 162.
32 Anonymous, *A History of Wonderful Inventions*, 1865, p. v.
33 *Ibid.*, pp. vi–vii.

Chapter 4: From Jules Verne to H. G. Wells

1 Winwood Reade, *The Martyrdom of Man*, 1872, p. 512.
2 *Ibid.*, p. 515.
3 Some of their publications were: George Griffith, *The Angel of the Revolution*, 1893; *The Outlaws of the Air*, 1895; *The Great Pirate Syndicate*, 1899; *The World Masters*, 1903; Robert Cromie, *For England's Sake*, 1889; *The Next Crusade*, 1896; *A New Messiah*, 1902; Fred T. Jane, *Blake of the 'Rattlesnake'*, 1895; *To Venus in Five Seconds*, 1897; *The Violet Flame*, 1899; William LeQueux, *The Great War in England*, 1894; *The Invasion of 1910*, 1906; *The Unknown Tomorrow*, 1910; Louis Tracy, *The Final War*, 1896; *An American Emperor*, 1897; *The Lost Provinces*, 1898.
4 Robert Cromie, *The Crack of Doom*, 1895, pp. 108–9.
5 Louis Tracy, *The Final War*, 1896, p. 372.
6 William Delisle Hay, *Three Hundred Years Hence*, 1881, p. 351.
7 Sir Julius Vogel, *Anno Domini 2000*, 1889, p. 96.

8 H. G. Wells, 'The Discovery of the Future', 1902, p. 92.
9 H. G. Wells, *A Modern Utopia*, 1905, p. 18.

Chapter 5: Ideal States and Industrial Harmonies

1 Thomas Erskine, *Armata*, 1817, Part 1, p. 168.
2 Anonymous, *New Britain*, 1820, p. 242.
3 Edward Bulwer-Lytton, *The Coming Race* (1871), Knebworth Edition, 1910, p. 232.
4 R. P. (Richard Walker), *Oxford in 1888, a Fragmentary Dream by a Sub-Utopian*, 1838, pp. 60–1.
5 *Ibid.*, p. 30.
6 Alphonse de Lamartine, *France and England: a Vision of the Future*, 1848, p. 37.
7 Henri Comte de Saint-Simon, *Selected Writings*. Translated and edited by F. M. H. Markham, 1852, p. 68.
8 Alphonse de Lamartine, *op. cit.*, pp. 37–8.
9 Émile Souvestre, *Le Monde tel qu'il sera*, 1859, pp. 3–4.
10 Thomas Carlyle, 'Signs of the Times', *Essays*, 1872, v. 2, p. 233.
11 *Ibid.*, p. 245.
12 Mary Griffith, *Three Hundred Years Hence* (1836), 1950, p. 123.
13 Edmund Ruffin, *Anticipations of the Future to serve as Lessons for the Present*, 1860, p. 302.
14 *Ibid.*, p. 341.
15 Dioscorides (Pieter Harting), *Anno Domini 2071*. Translated from the Dutch ... by Dr Alex V. W. Bikkers, 1871, pp. 71–2.
16 *Ibid.*, pp. 72–3.
17 Le Docteur H. Mettais, *L'An 5865, ou Paris dans quatre mille ans*, 1865, p. 5.
18 Alexander Pope, *The Dunciad*, 1742, Book IV, lines 649–56.
19 *Annual Register*, 1871, Part 1, p. 358. The editor included the *Battle of Dorking* and *The Coming Race* amongst the books of the year 'which have made the widest impression upon the public'.
20 James Presley, *Notes and Queries*, 4S, XII, 1873, p. 22.
21 Ismar Thiusen (John MacNie), *Looking Forward, or the Diothas*, 1889, pp. iii–iv.
22 Ernst Haeckel, *The History of Creation*, (1867), 1876, v. 2, p. 367.

Chapter 6: The Best of Worlds and the Worst of Worlds

1 The Earl of Lytton, *The Life of Edward Bulwer-Lytton*, 1913, v. 2, p. 465.
2 Samuel Butler, *Erewhon* (1872), 1910, pp. 87–8.

3 H. G. Wells, *The Time Machine* (1895), 1949, pp. 45–6.
4 *Ibid.*, pp. 89–90.
5 Richard Jefferies, *After London* (1885), Everyman's Library, 1948, pp. 32–3.
6 W. H. Hudson, *A Crystal Age* (1887), 1919, pp. 293–4.
7 *The Oxford Book of Greek Verse in Translation*, edited by T. F. Higham and C. M. Bowra, 1953, 'When Justice dwelt on Earth', translated by George Allen, p. 572.
8 W. H. Hudson, *op. cit.*, p. 149.
9 Arthur Bennett, *The Dream of an Englishman*, 1893, p. 93.
10 Robert William Cole, *The Struggle for Empire*, 1900, p. 19.
11 David Goodman Croly, *Glimpses of the Future*, 1888, pp. 36–7. Croly was a journalist and for a time editor of the *Daily Graphic*. He wrote extensively in support of Comte's theories. Much of the material in his book came from articles for 'The Prophetic Department' in the *Record & Guide*—an early example of forecasting.
12 John Jacob Astor, *A Journey in Other Worlds: a Romance of the Future*, 1894, pp. 41–2.
13 Georges Sorel, *Reflections on Violence*. Translated by T. E. Hulme and Edward A. Shills, 1950, p. 142.
14 Ferdinand Tönnies, *Community and Association*. Translated by Charles P. Loomis, 1955, p. 88.
15 Quoted by Ben Fuson, 'A Poetic Precursor to Bellamy's Looking Backward' in Thomas D. Clareson (ed.), *SF: The Other Side of Realism*, 1971, pp. 281–8.
16 William Delisle Hay, *Three Hundred Years Hence*, 1881, p. 354.
17 *Ibid.*, p. 278.
18 *Ibid.*, p. 109.
19 *Ibid.*, p. 257.
20 Francis Darwin (ed.), *The Autobiography of Charles Darwin and Selected Letters*, 1958, p. 69.
21 W. D. Hay, *op. cit.*, p. 231.
22 *Ibid.*, p. 257.
23 See entry for Edward Bellamy in the *Dictionary of American Biography*.
24 Edward Bellamy, *Looking Backward, 2000–1887* (1888), 1962, p. 105.
25 Émile de Lavelaye, 'Two New Utopias', *Contemporary Review*, January 1890, p. 19.

G. K. Chesterton recalled that his Uncle Sidney was an early follower of Bellamy: 'I remember him assuring me quite eagerly of the hopeful thoughts aroused in him by the optimistic official prophecies of the book called *Looking Backward.*' (*Autobiography*, 1936, p. 25.)

26 C. F. C. Masterman, 'Realities at Home' in: C. F. C. Masterman (ed.), *The Heart of Empire*, 1901, p. 50.

27 May Morris, *William Morris*, 1936, v. 2, p. 459.

28 *Commonweal*, June 22nd, 1889.

29 Graham Wallas, *The Great Society*, 1923, p. 326.

30 H. G. Wells, *A Modern Utopia* (1905), 1925, pp. 37–8.

Chapter 7: Possibilities, Probabilities, and Predictions

1 Jeanlouis Cornuz (ed.), *Oeuvres complètes de Victor Hugo*, 1960–3, v. 33, p. 551.

2 Matthew Arnold, *Culture and Anarchy* (1869), 1960, p. 204.

3 J. R. Seeley, *The Expansion of England*, 1883, p. 1.

4 Herbert Spencer, *The Study of Sociology*, 1897, pp. 399–400.

5 Paul Meuriot, *Des Agglomérations urbaines dans l'Europe contemporaine: essai sur les causes, les conditions, les conséquences de leur développement*, 1898, p. 453.

6 *Hygeia – a City of Health.* A Presidential address delivered before the Health Department of the Social Science Association at the Brighton Meeting, October 1875, by Benjamin Ward Richardson, M.D., F.R.S., pp. 8–9.

7 *Ibid.*, pp. 10–11.

8 Ebenezer Howard, *Garden Cities of To-morrow*, edited with a preface by F. J. Osborn, 1946, p. 235.

9 *Ibid.*, p. 128.

10 *Ibid.*, p. 145.

11 *Ibid.*, p. 146.

12 Alphonse de Candolle, *Histoire des sciences et des savants depuis deux siècles*, 1873, pp. 297–8.

13 Colonel F. J. Maurice, *The Balance of Military Power in Europe*, 1888, p. xxxi.

14 Rear-Admiral P. Colomb, Colonel F. J. Maurice etc., *The Great War of 189–*, 1893, p. 84.

15 I. S. Bloch, *Is War now Impossible?*, 1899, p. xvi.

16 *Illustrated London News*, November 17th, 1889, p. 824.

17 Alfred Marshall, *Principles of Economics*, 1890, p. ix.

18 *Ibid.*, p. 729.

19 Beatrice Webb, *My Apprenticeship*, 1926, p. 122.

20 Benjamin Kidd, *The Control of the Tropics*, 1898, p. 2.

21 Charles Pearson, *National Life and Character*, 1893, p. 363.
22 *Ibid.*, p. 27.
23 Ernest Udny, *The Freeland Colony*, with a Preface by Dr Hertzka, 1894, pp. 39–40.
24 Bernard Shaw, 'The Illusions of Socialism' in: Edward Carpenter (ed.), *Forecasts of the Coming Century*, 1897, p. 171. Another indication of the contemporary interest in the future appeared in *Popular Science Monthly* (51, 1897, pp. 307–14) where William Baxter began an article on 'Forecasting the Progress of Invention' with the observation: 'The great progress made during the last fifty years in the domain of science and invention has aroused a very general desire among intelligent people to know what the future has in store, and in many cases the desire has become so strong as to develop prophetic tendencies. Whenever a banquet is given in commemoration of some scientific event, or upon the anniversary of some ancient and honourable society, the orator of the evening is sure to dwell at considerable length upon the great discoveries that are still to come.'
25 *Ibid.*, pp. 100–1.
26 Field-Marshal Viscount French, *1914*, 1919, p. 11.
27 Charles Richet, *Dans cent ans*, 1892, p. 156.
28 *Ibid.*, pp. 62–3.
29 Norman and Jean Mackenzie, *The Time Traveller*, 1973, pp. 297–300.

Chapter 8: The Exploration of the Future

1 H. G. Wells, *Anticipations of the Reaction of Mechanical and Scientific Progress upon Human Life and Thought*, 1902, pp. 260–1.
2 *Ibid.*, p. 276.
3 *Ibid.*, pp. 315–16.
4 *Ibid.*, pp. 137–8.
5 There were numerous translations of Herzl's pamphlet and of *Altneuland*. The first English translation of the pamphlet was: *A Jewish State; an attempt at a modern solution of the Jewish question*, translated by Sylvie d'Avigdor, 1896.
6 Émile Zola, *Fruitfulness*. Translated by E. A. Vizetelly, 1900, pp. 410–11.
7 Émile Zola, *Work*. Translated by E. A. Vizetelly, 1901, pp. 498–9.
8 Prince Kropotkin, 'The Ethical Need of the Present Day', *Nineteenth Century*, August 1904, p. 207.

9 Somerset Maugham, *A Writer's Notebook*, 1949, p. 207.
10 Sir Leslie Stephen, *History of English Thought in the Eighteenth Century*, 1902, v. 1, p. 3.
11 Osbert Sitwell, *Great Morning*, 1948, p. 229.
12 William Stanley, *The Case of The. Fox*, 1903, p. 28.
13 *Ibid.*, p. 13.
14 G. K. Chesterton, *The Napoleon of Notting Hill* (1904), 1950, p. 10.
15 H. G. Wells, *The Fate of Homo Sapiens*, 1939, pp. 8–9.
16 H. G. Wells, *The World Set Free*, 1914, p. 65.
17 H. G. Wells, *In the Days of the Comet* (1906), 1927, p. 89.
18 G. K. Chesterton, *Heretics*, 1905, p. 74.
19 Anonymous (Harold Monro and Maurice Browne), *Proposals for a Voluntary Nobility*, 1907, p. 7. See: Joy Grant, *Harold Monro and the Poetry Bookshop*, 1967.
20 *Ibid.*, p. 20.
21 *Cosmopolitan Magazine*, November 1908, p. 569.
22 W. C. D. Whetham, 'Matter and Electricity', *Quarterly Review*, January 1904, p. 126.
23 H. W. Hillman, *Looking Forward: the Phenomenal Progress of Electricity in 1912*, 1906, p. 5.
24 Karl Pearson, *Social Problems*, 1912, p. 4.
25 William Barry, 'Forecasts of Tomorrow', *Quarterly Review*, July 1908, pp. 1–27.
26 Oswald Spengler, *The Decline of the West*. Translated by Charles Francis Atkinson, 1939, p. 3.
27 John Freeman, *Poems New and Old*, 1920, p. 298. (By courtesy of Miss C. Freeman.)

Chapter 9: From Bad to Worse

1 Alfred Duff Cooper, *Old Men Forget*, 1953, p. 48.
2 Herbert Read, *The Contrary Experience* (1963), 1973, p. 210.
3 Robert Graves, *Goodbye to All That* (1929), 1960, p. 146.
4 Edward Shanks, *The People of the Ruins*, 1920, p. 111.
5 Cicely Hamilton, *Lest Ye Die*, 1922, p. 121.
6 J. J. Connington (A. W. Stewart), *Nordenholt's Million* (1923), 1946, p. 181.

The book contained the first account of the world-wide dissemination of bacteria by aircraft: 'I have described the ramifications of the great air-services; and it seems to me obvious that the organisms were carried to and fro upon the surface of the globe by the agency of the aeroplanes.' p. 38.

7 *Ibid.*, p. 183.
8 Bertrand Russell, *Icarus*, 1924, pp. 62–3.
9 J. L. Mitchell, *Gay Hunter*, 1934, p. 82. Also *Today and Tomorrow* series: *Hanno, or The Future of Exploration*, 1928.
10 Karel Capek, *R.U.R.*, 1921, p. 78.
11 Capek complained that in a discussion held in London on *R.U.R.* Shaw and Chesterton were more interested in the Robots than in men; and he published an answer, 'The Meaning of R.U.R.', in *The Saturday Review*, July 21st, 1923. See: William E. Harkins, *Karel Capek*, 1962.
12 Fyodor Dostoievsky, *The Brothers Karamazov*. Translated by David Magarshack, 1958, v. 1, p. 304.
13 Yevgeny Zamyatin, *We*. Translated by Bernard Guilbert Guerney, 1970, p. 126.
14 *Ibid.*, p. 85.
15 *Ibid.*, p. 259.
16 Ayn Rand, *Anthem*, 1938, pp. 138–9.
17 Sir Daniel Hall, *The Pace of Progress*, 1935, p. 7.
18 Olaf Stapledon, *Last and First Men. A Story of the Near and Far Future* (1930), 1934, p. v.
19 Oswald Spengler, *The Decline of the West*, 1939, v. 2, p. 507.
20 Some of the more interesting forecasts were: H. G. Carlill, *Socrates, or the Emancipation of Mankind*; Garet Garett, *Ouroboros, or the Mechanical Extension of Mankind*; H. S. Jennings, *Prometheus, or Biology and the Advancement of Man*; André Maurois, *The Next Chapter: the War against the Moon*; Oliver Stewart, *Aeolus, or the Future of the Flying Machine*; L. L. Whyte, *The Future of Physics.*
21 These began with Jason (*pseud.*), *Past and Future*, 1918; C. Ballod, *Der Zukunftstaat*, 1919; J. M. Keynes, *The Economic Consequences of the Peace*, 1920; Karl Pearson, *The Science of Man*, 1920; Walter Rathenau, *In Days to Come*, 1921 (*Von kommenden Dingen*, 1917); N. Hamada, *An Ideal World*, 1922; Ralph Adams Cram, *Towards the Great Peace*, 1923; Bertrand and Dora Russell, *The Prospects of Industrial Civilization*, 1923; A. M. Low, *The Future*, 1925; Henry Ford, *Today and Tomorrow*, 1926; Adolf Caspary, *Die Maschinenutopie*, 1927; A. E. Wiggam, *The Next Age of Man*, 1927; *The Drift of Civilization* (Articles from the 50th Anniversary Number of the *St Louis Post-Dispatch*, December 9th, 1928), 1929; Rear-Admiral Murray F. Sueter, *Airmen or Noahs*, 1928; Sigmund Freud, *The Future of an Illusion*, 1928; G. F. Wates, *All for the Golden Age*, 1928; H. G. Wells, *The Way the World Is Going*, 1928; Hugh Ferris, *The Metropolis*

of Tomorrow, 1929; C. M. Hattersley, *This Age of Plenty*, 1929; Georges Duhamel, *Scènes de la vie future*, 1930; H. G. Wells, *The Way to World Peace*, 1930; Charles A. Beard, *America Faces the Future*, 1931; H. G. Wells, *What Are We to Do with Our Lives?*, 1931; David Lasser, *The Conquest of Space*, 1932; Frank Lloyd Wright, *The Disappearing City*, 1932; C. C. Furnas, *America's Tomorrow*, 1932. Thereafter, during the 1930s, the output of forecasts increased. In 1933 there were: Frank Arkright, *The ABC of Technocracy*; L. A. Fenn, *The Project of a Planned World*; Graham A. Laing, *Towards Technocracy*; L. Urwick, *Management of Tomorrow*; A. P. Young, *Forward from Chaos*. In 1934 there were: Ritchie Calder, *The Birth of the Future*; Effendi, *The World of Tomorrow*; E. T. Fradkin, *The Air Menace*; Christian Gauss, *A Primer for Tomorrow*; George Soule, *The Coming American Revolution*. In 1935: Kenneth Ingram, *The Coming Civilization*; J. N. Leonard, *Tools of Tomorrow*; Alfred Norman, *The Best Is Yet to Be*; R. Steiner, *The Social Future*. In 1936: C. C. Furnas, *The Next Hundred Years*; Roy Helton, *Sold out to the Future*; John Langdon-Davies, *A Short History of the Future*; J. P. Lockhart-Mummery, *After Us*. In 1937: W. P. Dreaper, *The Future of Civilisation and Social Science*; Robert Sinclair, *Metropolitan Man: The Future of the English*. In 1938: L. Cranmer-Byng, *Tomorrow's Star*; C. Delisle Burns, *Civilisation: the Next Step*; F. S. Marvin, *The New Vision of Man*. And in 1939: Arnold Hahn, *Grenzenloser Optimismus*; Max Lerner, *It Is Later Than You Think*; H. G. Wells, *The Fate of Homo Sapiens*.

22 Another indication of the growing interest in forecasting appears in Julian Huxley's *If I Were Dictator*, 1934. In the introduction Huxley wrote: 'I have been for several years interested in planning. A great many of the ideas here put forward have had their origin in connexion with Political and Economic Planning (PEP), a private organization founded some years ago for the non-political discussion of planning problems.'

23 The Earl of Birkenhead, *The World in 2030 A.D.*, 1930, p. 15.

24 *Ibid.*, p. 20.

25 *Ibid.*, pp. 155–6.

26 *Ibid.*, pp. 132–3.

27 Aldous Huxley, *Brave New World Revisited*, 1964, p. 11.

28 Aldous Huxley, *Crome Yellow* (1921), 1925, p. 47.

29 *Ibid.*, p. 242.

30 Oswald Spengler, *Man and Technics*, 1932, p. 103.

31 Harold Nicolson, *Public Faces*, 1932, pp. 330–1.

32 Stuart A. Rice etc., *Report to the President's Research Committee on Social Trends … in the United States*, 1933, p. 3. The sociologist William F. Ogburn played a major role in the compilation of the report. Ogburn was a pioneer in the field of social statistics and an early practitioner in social forecasting. For further information see: Otis Dudley Duncan (ed.), *William F. Ogburn on Culture and Social Changes*, 1964.

Chapter 10: Science and Society, 1945–2001

1 H. G. Wells, *A Short History of the World* (1946), 1956, p. 349.
2 C. E. M. Joad, *Shaw*, 1949, p. 238.
3 Somerset Maugham, *A Writer's Notebook*, 1949, p. 270.
4 Simone Weil, *Selected Essays*. Translated by Richard Rees, 1962, p. 44.
5 Hermann Hesse, *The Glass Bead Game* (1943), 1970, pp. 12–13.
6 *Ibid.*, p. 233.
7 W. H. Auden, *New Year Letter*, 1941, lines 1003–33.
8 W. H. Auden, *For the Time Being*, 1945, pp. 90–1.
9 Aldous Huxley, *Brave New World Revisited*, 1964, p. 12.
10 Pitirim A. Sorokin, *Social and Cultural Dynamics* (1941), 1962, v. 4, p. v.
11 Arnold Toynbee, *A Study of History* (1954), 1969, v. ix, p. 757.
12 Edwin Muir, *The Story and the Fable*, 1940, p. 257.
13 Aldous Huxley, *Ape and Essence*, 1949, p. 96.
14 George Orwell, *Nineteen Eighty-Four* (1949), 1972, p. 241.
15 *Ibid.*, p. 26.
16 *Ibid.*, p. 214.
17 Robert Graves, *Seven Days in New Crete*, 1949, pp. 38–9.
18 *Ibid.*, p. 41.
19 Marshall McLuhan, *The Mechanical Bride* (1951), 1967, p. v.
20 Jean-Paul Sartre, *What is Literature?* Translated by Bernard Frechtman, 1967, p. 210.
21 Bernard Wolfe, *Limbo '90*, 1961, p. 152.
22 Isaac Asimov, *The Caves of Steel*, 1969, p. 22.
23 Brian Aldiss and Harry Harrison (eds), *Hell's Cartographers*, 1975, p. 158.
24 Jean-Paul Sartre, *op. cit.*, p. 174.
25 Anthony Burgess, *A Clockwork Orange*, 1962, p. 122.
26 B.B.C. statement, 1966.
27 Peter Porter, 'Your Attention Please', in *Seven Themes in Modern Verse*, 1968, edited by Maurice Wollman, p. 141.

Chapter 11: Into the Third Millennium

1 B. F. Skinner, *Walden Two* (1948), 1967, p. 168.
2 Grover Smith (ed.), *The Letters of Aldous Huxley*, 1969, p. 740.
3 Ursula Le Guin, *The Left Hand of Darkness*, 1973, p. 21.
4 Ursula Le Guin, *The Dispossessed*, 1974, p. 286.
5 Quoted by David L. Porter, 'The Politics of Le Guin's Opus', *Science Fiction Studies*, II, 3, November 1975, p. 247.
6 Brian Aldiss and Harry Harrison (eds), *Hell's Cartographers*, 1975, p. 189.
7 John M. Mansfield, *Man on the Moon*, 1969, p. 201.
8 Pierre Teilhard de Chardin, *The Phenomenon of Man*, with an introduction by Sir Julian Huxley, 1959, p. 28.
9 Ossip K. Flechtheim, *History and Futurology*, 1966, p. 64.
10 Ward Moore, *Greener Than You Think*, 1947, p. 318.
11 Professor Joshua Lederberg, 'A Crisis in Evolution' in: *The World in 1984*, edited by Nigel Calder, 1965, v. 1, p. 28.
12 Brian Aldiss and Harry Harrison (eds), *op. cit.*, p. 92.
13 Stanislaw Lem, 'Philip K. Dick: A Visionary among the Charlatans', in *Science Fiction Studies*, II, 1, March 1975, pp. 54–5.
14 *Ibid.*, p. 55.
15 Stanislaw Lem, *Fantastyka i Futurologia*, 1973, v. 2, p. 43.
16 C. H. Waddington, *Operational Research in World War 2*, 1973, p. viii.
17 *Ibid.*, p. ix.
18 *Ciba Foundation Symposium 36* (New Series), *The Future as An Academic Discipline*, 1975, p. 1.
19 Colin Clark, *The Economics of 1960*, 1942, p. ix.
20 Sir Charles Galton Darwin, *The Next Million Years*, 1952, p. 20.
21 *Ibid.*, p. 135.
22 Harrison Brown, *The Challenge of Man's Future*, 1954, p. 266.
23 Sir George Thomson, *The Foreseeable Future*, 1957, p. 120.
24 *Ibid.*, p. 145.
25 Professor Gerard O'Neill, 'A Lagrangean Community?', *Nature*, v. 250, August 23rd, 1974, p. 636.
26 Nigel Calder, *The World in 1984*, 1964, v. 1, p. 8.
27 Hermann Kahn and Anthony J. Wiener, *The Year 2000*, 1967, p. xxv.
28 Michael Young (ed.), *Forecasting and the Social Sciences*, 1968, p. 54.
29 Pierre Teilhard de Chardin, *The Future of Man*, 1964, p. 306.
30 Alfred Tennyson, 'The Lotos Eaters', iv, lines 10–15, first published in *Poems*, 1832.

Index